D1253500

# GNAR COUNTRY

# Also by Steven Kotler

*The Angle Quickest for Flight*

*West of Jesus*

*A Small Furry Prayer*

*Abundance*

*The Rise of Superman*

*Bold*

*Tomorrowland*

*Stealing Fire*

*Last Tango in Cyberspace*

*Mapping Cloud Nine*

*The Future Is Faster Than You Think*

*The Art of Impossible*

*The Devil's Dictionary*

**Gnar:** adjective, short for "gnarly"; def.: any environment or situation that is high in perceived risk and high in actual risk.

**Country:** noun; def.: any defined territory, landscape, or terrain, fictitious or real.

# GNAR COUNTRY

## GROWING OLD, STAYING RAD

STEVEN KOTLER

HARPER WAVE
*An Imprint of* HarperCollins*Publishers*

 For information, address HarperCollins Publishers, 195 Broadway, New York, NY 10007.

HarperCollins books may be purchased for educational, business, or sales promotional use. For information, please email the Special Markets Department at SPsales@harpercollins.com.

All art provided by author with the exception of page iv: Michal Durinik/ Shutterstock, Inc.

FIRST EDITION

Library of Congress Cataloging-in-Publication Data has been applied for.

ISBN 978-0-06-327290-3

23 24 25 26 27 LBC 5 4 3 2 1

For the Trees

To do a dangerous thing with style is what I call art.

—CHARLES BUKOWSKI

# Preface

According to Wikipedia, "punk rock (or simply 'punk') is a music genre that emerged in the mid-1970s. Rooted in 1960s garage rock, punk bands rejected the perceived excesses of mainstream 1970s rock. They typically produced short, fast-paced songs with hard-edged melodies and singing styles, stripped-down instrumentation, and often shouted . . . anti-establishment lyrics. Punk embraces a DIY ethic. [For example], many bands self-produce records and distribute them through independent record labels."

The most punk rock thing about me? I learned to ski on converted garbage dumps. That was in Cleveland, Ohio. That was in the 1970s.

That was a long time ago.

# Introduction

This is a book about goals and grit and progression, especially in the second half of our lives. It's a book about what it takes to fight off that quaint human urge to die a little bit each day. It's an antidote for the weariness.

Don't think of it as a "how-to" book. Think of it as a "how-not-to" book. How not to lose that brash fire. How not to give in to that cozy blanket of middle age. How not to go gently into that good night.

In short, this is a book about growing old and staying rad. In even shorter, this is a book about skiing.

Of course, if you're wondering what punk rock and skiing have to do with growing old and staying rad—well, good question. Yet to answer it, I need to tell you a little bit more about where this book came from, and that requires telling you a story about skiing.

So, for now, back to the skiing.

## KIRKWOOD, FEBRUARY 26, 2020

According to *The Liftie Report,* Kirkwood's Oops and Poops is the tenth-steepest run in Tahoe. Of course, you won't find Oops and

Poops on a standard map of Kirkwood. On the map, you'll see the double black diamond Chamoix perched above a thick glade of trees with the words "avalanche boundary" printed across them.

But here's one way to tell the tourists from the locals: If you go to a resort, ride the lift, and ski a run that's clearly identifiable by a sign with a name—you're probably a tourist. If you ride the top of run A into the little chute in the trees separating run A from run B, then zip across run B to hit the little cliff that launches you onto run C, and so forth across the mountain—you're arguably a local. Locals like to draw unusual lines across their mountains. Tourists usually follow the lines the ski area has predrawn for them.

Oops and Poops is one of those lines that locals have drawn. It starts just right of Chamoix, with a nosebleed-steep, cliff-lined face that launches skiers into a sneaker chute through the forest that ends at a keyhole notch formed by a large rock wedged between tall pines. Dropping through the keyhole puts you atop a ten-foot-wide natural halfpipe that descends for over a thousand feet of crazy angled possibilities. Think pinball, at thirty miles per hour.

Riding the pipe well requires a significant amount of creativity. The tourist choice is the center line: Just ski down the middle of the damn pipe and be done with it. But the middle of the pipe is actually strewn with boulders that form mini-cliffs at often inconvenient moments. These obstacles may result in that familiar mountain experience known as "shitting your pants," thus, the run's name: Oops and Poops.

Plus, on a personal note, I think the center line is bullshit. Most tourists don't ski ten-foot-wide, cliff-lined corridors with ease, thus their choppy, staccato, weak-ass attempts at turning create a drunken array of asymmetrical moguls that don't mix with my flow-and-destroy approach to skiing.

I'd also like to mention that "tourist" is the polite term. In the parlance of mountain culture, a tourist is a "Jerry." As Thacher Stone, creator of the Jerry of the Day video series, once explained to *Freeskier* magazine: "Within the ski industry, a 'Jerry,' otherwise known as a 'Gaper,' a 'Joey,' a 'Gorb,' etc., is someone doing a boneheaded move. My personal definition of a Jerry is: 'An individual who exhibits a true lack of understanding for their sport, or for life in general.'"

Locals, on the other hand, approach their sport with high-speed creativity, a form of intuitive problem solving that I've come to think of as "fast geometry." Oops is a solid example. If you want to ride Oops with style and flow, beware the lure of the center line. Instead, with gleeful and wanton abandon, hurl your meat carcass onto the walls of the pipe for a round of fast geometry.

The walls are steep, running anywhere from thirty-five degrees to dead vertical. At most ski areas, a tilt of thirty-five to forty degrees is a single black diamond run of the "most difficult" variety. Double black diamonds—"experts only"—fall away at forty to fifty-five degrees, or steep enough that you can reach out your hand and touch the slope. Above forty-five degrees, any turn you make will likely include a three-foot freefall before you reencounter the surface of the Earth, which explains the "hop-and-drop" technique developed by early steep skiing pioneers.

The secret to riding a line like Oops creatively is to ride the entire pipe. Make big arcing turns onto the wall, hop off something—usually a stump or rock—and drop into the smooth transition that blends the pipe's wall with its bottom and with enough acceleration to ride up the other side of the wall and repeat the process. Of course, in a normal halfpipe, the walls aren't lined with obstacles, and the pipe doesn't take abrupt, forty-five-degree turns, as the Oops's pipe does on three different occasions—but whatever.

When you ski a run like Oops at speed, which is the only way to ski this line with any fluidity, you don't get to see much. All the brain really gets is a millisecond to recognize a terrain feature, recognize the feature's angles, recognize the trajectory that the human body will fly off those angles, and then it's go time. Fast geometry. Compute, then execute or tomahawk. What does "tomahawk" mean? It means to tumble, ass over teakettle, down the side of a mountain.

On February 26, 2020, Ryan Wickes, my ski partner, and I were skiing Oops on the last powder day of the season. We didn't know it was the last powder day, nor was there much powder. Maybe two inches of new snow had fallen. But we hadn't seen new snow since January and the presence of the fluff brought out the rowdy.

I remember little from the day except blazing down Chamoix, carving hard right under the cliff, and blasting through the keyhole notch and onto the top of the Oops's halfpipe. I leapt off something, bounded over something else, then carved high on the pipe's right wall. About ten feet in front of me was a little pillow of fresh snow, perched atop a small boulder beside a towering pine. My brain saw the snow pillow and made an unusual suggestion: sliding spin 360.

The thing about fast geometry—there's no time for internal debate. When your brain makes a suggestion, you either execute immediately or drop out of flow because you're now catawampus and taking an awkward line down the mountain. Or, you're catawampus, taking an awkward line down the mountain, and about to spend a less-than-lovely evening in the nearest emergency room.

Still, a suggestion like "sliding spin 360" was the kind of thing that deserved discussion. For starters, I'd never thrown a sliding spin 360 on the wall of a halfpipe before, and definitely not on a wall chockablock with obstacles. Additionally, I was doing close to thirty miles per hour and this particular section of wall was close to forty degrees steep. Misjudge the angle of the spin and hit the tree. Misjudge the velocity of the spin and there's a long fall to the dead flat of the pipe's bottom to consider your error.

What is a sliding spin 360? We'll get there. For now, know that it's a trick that any talented ten-year-old skier can throw, and most stop throwing them before they reach age eleven. But I wasn't ten. I was fifty-three.

Fuck it. Here in the twenty-first century, fifty-three is the new ten. I threw the damn sliding spin. Gnar Country for Old Men. Game on.

### NORTHERN NEVADA, FEBRUARY 27, 2020

Almost nailed the sliding spin. I got three-quarters of the way around before the snow stole my speed. Instead of a 360, I spun a 270. But I had a good angle off the pillow, so I hopped out of danger and into the halfpipe and, from somewhere behind me, Ryan cheered.

In nearly a decade of skiing together, I'd never heard Ryan raise his voice, not even when dangerous shit goes wrong. Hearing Ryan cheer—unusual, to say the least.

I got home from Kirkwood exhausted, but went to sleep dreaming of possibility. I woke up to an entirely different possibility. I was running a high fever, even more exhausted than before, and, oddly, unable to taste food.

By the time I could breathe again . . .

### KIRKWOOD, MARCH 14, 2020

The day the season officially died. COVID infection rates were thirty times higher in ski towns. Today the governor of Colorado pulled the plug on the resorts. Every state in the nation followed suit. My skier's soul was dashed upon the rocks, shattered into a million pieces, and nothing could put Humpty Dumpty back together again.

### NORTHERN NEVADA, MARCH 15, 2020

Early days of the lockdown. I watched SLVSH videos on YouTube. Nothing else made sense.

**NORTHERN NEVADA, MARCH 18, 2020**

Still watching SLVSH videos—this might take some explaining.

SLVSH—pronounced "slush"—is the freestyle skiing version of the basketball game HORSE. Two skiers serve as competitors. A quick round of rock paper scissors decides who goes first. The winner calls out a trick—complete with the name of the trick, the name of the grab they're going to execute during the trick, and the location where they're going to throw the trick. For example, "A rodeo 540 safety off the first kicker."

In this case, a "rodeo" is an off-axis flip thrown backward. If you're standing on flat ground, leap into the air, and kick both your feet over your left shoulder—thus, flipping sideways and backwards—that would be a rodeo. The "540" refers to the degrees of spinning rotation. A 540 means you take off forward, spin one and a half circles—or 540 degrees of rotation—and land backwards. The "safety" refers to the type of grab the skier throws while flipping and spinning, with a "safety" meaning the skier grabs the part of their ski directly under their boot.

In SLVSH, after calling out the trick's name, the first skier tries to execute the trick. If they pull it off, skier two must do the same. Miss a trick, get a letter. The letters spell SLVSH because, in the spring and summer conditions when the game is typically played, landings are soft and *slushy,* and skiers can try more dangerous and ambitious tricks.

All matches are captured on video, and then posted to SLVSH .com. Anyone can watch. What makes watching worthwhile is that the competitors are among the very best freestyle skiers in the world. It's like NBA stars Steph Curry and James Harden squared off for a playground game of HORSE, except, instead of shooting from half court, they were throwing double superman front flips over sixty-foot gap jumps.

## PINE NUT MOUNTAINS, MARCH 20, 2020

I took the dogs for a hike. We went uphill for a while, then we went downhill for a while. Somewhere in between, I lost my mind.

I had been obsessing over my sliding spin 360. I had been obsessing over SLVSH videos. As we hiked, I found myself contemplating a seriously dumb idea—learning how to park ski.

I knew almost nothing about park skiing. My history was as a "big mountain" skier. It's like the difference between speed skating and figure skating. Speed skating takes place in two dimensions; figure skating takes place in three.

As a big mountain skier, I was a speed skater. Most everything I did occurred while moving in a single direction and staying in contact with the surface of the snow. But that's not how park skiing works.

"Freestyle"—which is the technical term for the discipline that includes park skiing—often takes place above the surface of the Earth. It means skiing backwards and sideways and upside down. Most crucially, it involves tricks.

Since I wasn't a park skier, I had no idea how to do tricks. In fact, after nearly fifty years of skiing, my entire bag of tricks included a spread eagle, a backscratcher, a mule kick, and that sliding spin 360 I'd thrown at Kirkwood. As mentioned, a sliding spin is a trick learned by ten-year-olds and then quickly forgotten. A spread eagle, backscratcher, or mule kick? Unless you're going for retro chic, they're already forgotten. No one throws those tricks anymore. Not for decades.

Nor did I really throw those tricks anymore. I've had more season-ending injuries than I can count. You know the worst part about surgery? The lights. When they wheel you from pre-op into the operating theater, you're flat on your back on a gurney, staring up at those shitty fluorescent lights that line hospital ceilings. The thing about pain, the brain remembers every detail that precedes it, hardwiring the moment into memory so the next time this situation arises, you get the hell outta Dodge before the pain says hello. Man, I never wanted to see those lights again.

But that sliding spin 360 had unlocked a door, and those SLVSH videos made me very curious about what was on the other side.

# Chapter 1

Of course, I had to walk through that door. Of course, Gnar Country was what I found on the other side. Of course, one of the main things found in Gnar Country was skiing. More specifically, "freestyle skiing," aka park skiing, aka the branch of the sport devoted to tricks like the sliding spin 360 and the rodeo 540.

Yet this raises the question: Why join a fifty-three-year-old man on his quest to learn to park ski—especially if you're not interested in skiing?

First, there's comedy value.

According to traditional learning theories, it's supposed to be impossible for someone my age to learn to park ski. And I'm not talking about me reaching expert-level X Games status. For a half-dozen biological factors that we'll get to later, it's highly unlikely that anyone over the age of thirty-five can become even an intermediate park skier.

But not so fast.

Recent discoveries in embodied cognition, flow science, and network neuroscience have revolutionized how we think about human learning. On paper, these discoveries "should" allow older athletes to progress in supposedly "impossible" activities like park skiing. To see if theory worked in practice, I put these ideas to the test on the ski hill, conducting my own ass-on-the-line experiment in applied neuroscience and later-in-life skill acquisition—aka I tried to teach this old dog some new tricks.

That's how this book began.

Early on, I started keeping a performance journal—a record of what worked, what didn't, and why. Ten months into the experiment, I was having a conversation with my closest friend and longtime editor, Michael Wharton. I'd learned a new ski trick. Michael had a few questions about my process. I couldn't remember the exact details, but I told Michael about the performance journal. We dug it out to find those details.

Michael read that entry, then a few more entries, then a few more. "This should be a book," he said afterward.

No one was more surprised than me. As far as I knew, I was having a very private conversation with myself. Michael pointed out that my conversation was also a blow-by-blow account of what it looks like when an ordinary person uses the tools of peak performance to tackle an extraordinary challenge. In other words, as Michael said, "It's a written record of how the magic trick gets done."

We got off the phone. I reread the journal. I realized Michael was right. And since he's been my editor for twenty-five years, you have to know how much it pains me to admit that.

Still, while I've written six books on flow and peak performance, all of them are essentially introductory texts. That was Michael's point—as there was no other way to write those books. Because of the complicated nature of flow science and my desire for readers to understand that science, those books could only capture a small portion of what it actually means to apply these ideas on a day-to-day basis.

That wasn't the case with my journal. In fact, my rough, raw record of my season had managed to do something I couldn't manage to do in my last book, *The Art of Impossible*—which is ironic, since the thing I couldn't do in that book was to properly explain the title. *The Art of Impossible* is about the science of peak performance and how to use that science to tackle hard challenges. So why is the word "art" in the title?

The science of peak performance is a science. The application of that science in our day-to-day lives is an art. It's a wildly creative act that's different for everyone. In the face of real-world scary shit, using the full suite of peak-performance tools to rise to that challenge—and without needing to stop and define each of those tools—that's what my journal captured.

As a result, this book goes far beyond the basics of peak performance. It's a nuanced, expert-level view. It has less science than my other books—that's been saved for the appendix and the endnotes—and far more practice.

Yet the only way to communicate that view is to focus on a single "impossible" challenge in a discipline where I have expert-level skills—aka skiing—so I can actually explain what I'm talking about. But, so there's something to learn, I have to move into an adjacent activity—from big mountain skiing into park skiing—which is a realm where I am an absolute beginner, with zero experience and, let's be honest, a ton of fear.

And this was Michael's other point: this situation—me, trying to learn to park ski, in middle age, in the middle of a pandemic, with all the real-world responsibilities and fragilities those realities bring, using every peak-performance tool at my disposal—those things don't happen every day.

Yet unless you're willing to learn more about skiing than you might desire, they can't happen at all.

Chances are, if you're reading this, you're over the age of fifty or hope to be over the age of fifty sooner or later. This is a good news/bad news kind of thing. The good news is, "over fifty" doesn't mean what it used to mean. The bad news is, to explain why, you're gonna have to read about skiing.

Let's start with a high-level view.

Over the past decade, scientists have learned a great deal about maintaining vitality into our later years. Research into communities with exceptional longevity—technically known as Blue Zones—reveal five keys to a long, happy life: move around a lot, de-stress regularly, have robust social ties, eat well—meaning, eat mostly plants and not too much of anything—and try to live with passion, purpose, and regular access to flow. In other words, we now know the basic requirements for long-term health and well-being.

This isn't a book about those basics. Nor is it a book about their cousin, longevity science. Yet here too the news is good. In recent years, scientists have uncovered the nine major causes of aging. There are now billions of dollars and dozens of biotech companies aimed at eliminating each of them. In fact, thanks to continual advancements in biotech, every day we manage to stay alive, we gain an additional five hours of life expectancy.

The upshot of all this?

Most of us are going to be "old"—whatever the hell that means—for a lot longer than our ancestors and, quite possibly, a lot longer than we ever expected. So how do you want to spend that extra time? That is the topic of this book.

Put differently, if you believe you've already done "enough," this book isn't for you. If you're satisfied with your life and willing to ride into the twilight without chasing down one more magnificent dream, this book isn't for you. If you're just fine with long days of safe, secure, mediocre, average, golf, tennis, Top 40, crack a beer and watch a ball game, this book isn't for you.

But if you've ever asked: "What if?"

Yeah, this book is probably for you.

Thus, even if you don't like skiing, the sport remains the best way I know to teach you about applying the science of peak performance to the grown-up problem of "what if?"

And "grown-up" is more than a phrase. Later in life is when most of us have both the time and the resources for "what if?" Also, because we've finally lived long enough to know a thing, it's when our biggest dreams come into their clearest focus.

This is the best news.

A growing pile of research shows we can sustain peak performance further into life than anyone thought possible. Sure, physical skills begin to decline in our thirties. Sure, they drop roughly one percent a year thereafter. That's where the biology stands today. But we are also starting to understand how to offset this decline without a significant amount of performance loss. And there's an opportunity for enormous performance gain.

As we enter our fifties, if we get "it" right, we gain access to a suite of legitimate superpowers.

Over the course of that decade, there are fundamental shifts in how the brain processes information. In simple terms, our ego starts to quiet and our perspective starts to widen. Whole new levels of intelligence, creativity, empathy, and wisdom open up. As a result, key downstream skills like critical thinking, problem solving, creativity, communication, cooperation, and collaboration all have the potential—if properly cultivated—to skyrocket in our later years.

What's the big deal? I've just listed the six skills that experts agree are most crucial for thriving in the twenty-first century. More critical here, they're also the six skills that may make answering "what if?" a hell of a lot easier than expected.

Yet access to our superpowers isn't automatic.

To produce the necessary brain shifts, we must first pass through a series of gateways. Work in developmental psychology shows that crossing these thresholds is necessary for maintaining happiness and well-being into our august decades. More important here, if the goal is peak-performance aging, passing through these thresholds and unlocking our superpowers significantly helps offset the natural decline that comes with time.

What are those gateways?

By age thirty, we need to have figured out who we are in this world, solving the crisis of identity. By age forty, we have to figure out how to make a living that is seriously aligned with our big five intrinsic motivators: curiosity, passion, purpose, autonomy, and mastery. By fifty, we need to forget old grudges, forgive those who have done us wrong, and generally clear our emotional slate. Finally, somewhere along the way, we need to counteract the rising risk aversion and general fragility that accompany aging.

How to utilize the tools of peak performance to pass through these gateways, unlock our superpowers, and gleefully attack hard challenges throughout our later years—that is the topic of this book.

And this brings us to its title: *Gnar Country*.

The word "gnar" is short for "gnarly," which is action-sports slang for a very specific kind of danger. When athletes use the word, they're referring to situations that are high in perceived risk and high in actual risk. If I describe a ski line as "the gnar," I am telling you, if you're interested in attempting this line, you need both the physical skills to navigate the terrain and the mental skills to navigate the terror that comes with navigating the terrain.

"Country," as I'm using the term, is any defined territory or landscape. For example, our later years are a "country." They're an unknown territory, full of danger, full of surprises, and—as we'll see throughout this book—full of outdated notions about what is possible for each of us during our later years.

"Gnar Country," then, describes both the terrain of our later years—high in perceived risk, high in actual risk—and the gritty mindset needed to thrive during those years. Yet there's no way to teach you how to thrive during those later years without concrete examples, and this is another reason why skiing is useful.

For the past three decades, the ski mountain has been my personal laboratory. It's where I first test all my peak-performance ideas. If there's an interesting theory in a research paper, I put it into practice while skiing. If something works for me on the slopes, I'll bring it into the Flow Research Collective, launch a research initiative, and figure out if it works for other people.

In this book, I'm inviting you inside my laboratory for an experiment in high-stakes dream chasing. It's a personal journey into the practical side of the still somewhat uncharted realm of peak-performance aging. For this reason, *Gnar Country* is part adventure story, part living experiment, and part late-in-life peak-performance primer. Which is to say, if you're willing to read about some skiing, I'm willing to guarantee you'll come out the other side knowing how to kick ass until you kick the bucket.

Additionally, if you're unfamiliar with the basics of flow science and peak performance—that is, topics covered in my other books—*The Art of Impossible* is the best place to start. If you're not interested in reading another book, you can also go to www.gnarcountry.com and find a series of videos that cover the important ideas. You'll also find videos that cover the basics of freestyle skiing and, if you're unfamiliar with the sport, they'll give you visuals to go with my words and that will help.

One quick note about the structure of the book. My aim is to show and not tell. To really explore the practice of peak performance and peak-performance aging, we'll go day-by-day through my ski season, from preseason training through my last day on the snow. In these sections, I'll explain basic principles and key details. Toward the end of the journey, I'll reexamine these ideas, translating them out of skiing and into a set of general principles that you can use to tackle hard challenges later in life.

You're also going to learn a bit about punk rock in this book. Why? Two reasons. First, punk had an enormous influence on action sports. Second, punk had an enormous influence on me. But all this will become clearer as we go along.

For now, back to the skiing.

## NORTHERN NEVADA, MARCH 22, 2020

But there was no skiing. The resorts remained COVID closed, and I was not taking it well.

For reasons that we could file under "real life," I had worked nine years with nary a break. I was burned-out beyond belief. But I had a plan to deal with my burnout—and that was the real source of my anger.

For the past two years, I'd been preparing for the glory of April 2020. That month was going to be my break: a month devoted to skiing the Sierra Nevada Mountains. I had lived in these mountains back in the 1990s, on the California side. Now, in the 2020s, I had moved back to these mountains, on the Nevada side. However, I'd only just arrived.

I only got a handful of days on the snow prior to resort closure. It was a quick peck on the cheek; I wanted the full, torrid affair. Plus, the winter had been bone dry, for both snow and flow. There had been only two powder days, including the day I attempted that sliding spin 360. In fact, by the time I'd shaken off the jitters that came from learning my way around new mountains, they'd shuttered the mountains. Then there was no flow at all.

I was not prepared for the soul-crushing sadness. I was not prepared for the seething rage. I understood that there was nothing sane about these feelings. We were in the middle of a pandemic and the entire planet was suffering and I gave a shit that the ski mountains were closed?

But I did.

## NORTHERN NEVADA, MARCH 23, 2020

I called Ryan and mentioned that I couldn't do anything but watch SLVSH videos. I also mentioned that nobody I saw in those videos skied like me. As body position is tightly linked to embodied cognition, if I couldn't find anyone who skied like me, there was no practical way for me to model their skiing.

Ryan mentioned that, style-wise, I skied like a guy named Adam Delorme.

I said: "Really? On the pro meter?"

On the pro meter, explained Ryan, if Delorme was a ten, on a really good day, I was "maybe a three."

A three? *Maybe* a three? "Maybe a three" is the nicest thing anyone has ever said to me. I could work with "maybe a three."

Then, I spent the rest of the afternoon on the internet trying to figure out who Adam Delorme was.

## SKY TAVERN SKI AREA, MARCH 25, 2020

Magic March delivered. The heavens opened up and blessed the land with two feet of pow-pow.

With the resorts still COVID closed, Ryan and I decided to hike Sky Tavern, the tiny public ski hill outside of Reno. It wasn't quite as small as the garbage dumps converted into ski resorts that littered my Cleveland childhood, but they shared the same neighborhood. Still, there was a double black diamond run that led straight into the parking lot. I even brought out the brand-new fat skis, the Line Pescados.

So much for good intentions. We hiked to the top and I tried to sliding spin the Pescados right out the gate. No joy

I got grabbed by the pow-pow and knocked on my ass. It got worse from there. I was still recovering from COVID and could barely breathe. Our three laps nearly killed me. Plus, I hated hiking uphill to ski downhill. I'm in it for the downhill. I even have a T-shirt: "Life is too short to go uphill."

I'd also like to mention that my T-shirt is blasphemy in certain circles. In modern mountain culture, resort skiers are barely considered real skiers. Today, real skiers earn their turns in the backcountry. "Skin up, ski down" is the mantra of the modern adventure athlete.

I have nothing against this mantra. I have nothing against the backcountry. I spend a good portion of my time out there with my dogs. But the only time I hike up mountains to ski down mountains is after the resorts close and I can't catch a faster ride.

Some of this is practical. For me, skiing is usually a solo mission and "suicide" is the word skiers use to describe solo missions in the backcountry. More to the point, for me, skiing is actually a mission.

The first time I chased professional skiers around a mountain was in 1994. I was a journalist; action sports were my beat. Before then, I thought I was an expert skier. I started skiing at age eight, ski-bummed around Colorado after college, and could generally slide down any "experts-only" double black diamond run without making a mess.

In short, before skiing with pros, I thought I could keep up with pros.

I couldn't keep up. Not even close. The facts speak for themselves: It was 1994 when I first started chasing professional skiers around mountains. This was in Chamonix, France.

In 2013, I was in Jackson Hole, Wyoming. I was skiing with an all-star posse in full blaze mode when I noticed something unusual: We were all arriving at the chairlift at the same time. No one was waiting for me. After nineteen years of trying, I could finally keep up.

But I wanted to do more than keep up. I wanted what those pros had: the ability to use their sport for stylish self-expression. I wanted to play fast geometry with flair. If I saw a feature on the mountain that seemed interesting, then I wanted the ability to do something interesting with that feature. Art in, art out—the way it's supposed to work.

Mine was a desire that required practice. In the backcountry, you don't get enough laps for the kind of practice I required. Three or four laps is a good day in the backcountry. At the resort, it's twenty-five to thirty.

I needed the practice as I have little natural talent. As I child, I was skinny, klutzy, often slow, always scared. I spent grade school being the last guy picked for the team in almost every athletic contest where team picking was required. I was also the first guy the jocks liked to pick on, which meant I spent a lot of my childhood losing fights to jocks. Why did I become a punk rocker? One reason: the punks fucking hated the jocks.

Neither am I a natural action-sports athlete. Naturally, what I am is acrophobic. Heights provoke terror, terror provokes the freezing response. I spent a decade rock climbing in the hope that exposure to exposure would cure my fear. "Fail" is the operative term here.

Then a long bout with Lyme disease screwed up my inner ears and left me prone to vertigo. These days, whenever I encounter heights—unless I'm regularly encountering heights—first comes the fear, next comes the freezing, then the spinning, finally the leaving. First, courage deserts the scene, with coordination not far behind. Muscle control and motor dexterity depart next, with self-esteem in tow.

In the end, all that remains is the shame.

"Steazy" is the word that good skiers use to describe good skiing. It's a mash-up term, a blend of "stylish" and "easy," as in: "That dude is steazy in the sick gnar." If my goal was to be steazy in the sick gnar, you better believe I needed the practice.

Man, did I miss the resorts.

## NORTHERN NEVADA, MARCH 28, 2020

My seething rage at the shuttered ski resorts was growing dangerously worse. I was angry at the world. I was angry at God. What would make this okay? That was the question I kept asking myself.

I mean what, beyond climbing the mountain behind my house and screaming at the sky: "How do you spell 'God'? That's F-U-C-K-W-A-D."

Because I tried that today—and it didn't work.

## NORTHERN NEVADA, MARCH 29, 2020

I woke up to the truth. COVID would never leave. The ski resorts would never reopen. And it will never snow again, ever, anywhere, for all eternity.

What would make this okay? That was still the question. I still didn't have an answer.

*Iv jIH ghotvam*?

Nope, not even after I translated the question into Klingon.

## NORTHERN NEVADA, MARCH 30, 2020

What would make it okay? This begat a different question: What had been stolen from me? I already knew the answer: Progress had been stolen.

I lost the last two and a half months of the ski season, and the part that stunk? The knowing stunk. I knew I was getting older. I knew my window to make serious progress as a skier was shrinking. And I knew there was so much left to do.

Why was there so much left to do? The short version: the jocks. The bullying. The psychological hangover from the jocks and the bullying. Simply put, you can't spend your childhood being the last guy picked for kickball, and then spend your early adulthood chasing professional athletes around mountains and being the last guy to arrive at the chairlift—aka the snow sports equivalent of being the last guy picked for kickball—without it leaving a mark.

Why was there so much left to do?

Because "dirty old shame" is the technical term for this mark, and skiing is where I had decided to settle the score and erase that mark.

I took all this knowledge with me to bed last night. The answer was waiting for me when I woke up this morning. If I entered next season a better skier than I ended last season—that might make this okay. If, somehow, with the snow melted away, the resorts shut down, and a pandemic raging, I could significantly progress my skiing without being able to ski—yeah, that answer seemed to quiet the rage.

Next question: How do you progress as a skier if you can't actually ski?

## PINE NUT MOUNTAINS, APRIL 1, 2020

In the SLVSH videos, skiers were sliding rails. I'd never slid a rail in my life. I'd never considered sliding a rail. The whole rail-sliding phenomenon took hold of skiing during my thirties, a decade when I lived by the ocean and spent my time surfing. Then I moved to New Mexico, where the better terrain parks weren't at the places I liked to ski.

The Santa Fe Ski Area was closest to my house and has some of the best tree skiing in America, so that was what I did when I was there. Taos was usually about hiking the West Basin for the chutes up top or the mile of bumps that was Al's Run. And then Wolf Creek, in southern Colorado, which was my other regular stop, combines the steeps and the trees in a way that happens at no other ski area. It's not for everyone, but it worked for me.

Give or take writing some books, this was how I spent my forties.

At no time during this period did the notion of learning to park ski ever enter my consciousness. Why would it? Everybody knows that park skiing, aka freestyle skiing, aka the kind of skiing that involves throwing tricks off jumps and on rails, is a young person's game. Actually, that's putting it mildly. If you're over the age of thirty, learning how to park ski is considered "extremely difficult." If you're over the age of forty, it moves into the "downright crazy" category. Once you hit fifty, "that's impossible" becomes the standard response.

The facts are the facts. You hit the ground hard while learning to park ski, that's a fact. Another fact is that older athletes take a lot longer to heal than younger athletes. Additionally, park skiing requires significant fine-motor performance and fast-twitch muscle responses, and both atrophy as we age. Finally, the acquisition of motor patterns, like the acquisition of language, is an innate process in children. But that window of opportunity closes long before middle age. As a result, most older athletes don't believe they're capable of onboarding the entirely new language of motor patterns required for park skiing. For these reasons, for others, there's a general philosophy among action-sports athletes: If you haven't learned to park ski by the time you're thirty—don't bother.

But watching SLVSH had taught me a few things—the names of the tricks for one. I can now tell a "back swap, front two" from a "disaster on, four out" from a "50–50 on, nollie 180 out." And if none of these words mean anything to you, well, you're not alone. Most skiers don't even know what they mean. But what's important here is embodied cognition. While watching those videos, I'd realized the same motion used in my sliding spin 360 was used by park skiers on rails, in a trick known as a "surface swap 360." And I didn't realize this intellectually. My body realized it somatically. Did this realization apply in the real world? That was an interesting question.

Yet there's a gap between "interesting, this stuff has piqued my curiosity" and "interesting, this is stuff I should try." Typically, fear fills that gap. Rails are made from polished steel, slicker than ice, and have a long history of putting even the best skiers on their ass. Unfortunately, as you age, the distance between your ass and the emergency room is a lot closer than it used to be—or so sayeth the voice in my head.

All of this changed on the first day of April. I was hiking with Kiko, my best friend and a one-hundred-and-twenty-pound Maremma sheepdog, past an abandoned gold mine in the Pine Nut Mountains. The mine was perched on a small plateau between three small hills. The miners had left behind dozens of tailing mounds, enormous piles of finely ground rock and dirt, so smooth and steep they looked skiable.

Were they skiable?

The biggest mound around had a thirty-five-degree incline and was maybe fifty feet high. This was steep enough to get going and tall enough for a few turns, but a few turns weren't going to produce the progression I was chasing.

A jump was a different story. There were plenty of high-angle downslopes upon which to land, and plenty of jump tricks I wanted to learn. Yet landing on dirt is still landing on dirt, and I'd already learned this lesson the hard way while mountain biking.

Perhaps not a jump. What about a rail?

You need a lot of speed for a jump, but a lot less speed for a rail. All that was required, I suspected, was a ten-foot tailing pile that descended onto flat ground. I looked around. Right, left, up, down—the piles I needed were everywhere.

Of course, I knew I couldn't ski the dirt itself. There would be too much friction unless I had tons of speed. But what if I laid cardboard boxes atop the dirt. Could I ski on cardboard?

Memory lane came to my rescue.

Years earlier, I'd written a story about pro mogul camps that took place on Mount Hood during the summer months. They had water ramps up there—big monsters that launched you into bigger pools of water. The in-run to that jump was coated in plastic grass and kept wet, a combination that produced a surface slick enough to ski. Sure, you couldn't turn on the stuff, but as long as you didn't move a muscle from the moment you slid onto the in-run to the moment you launched off the jump, everything worked just fine.

Next question: Could I lay plastic grass atop these tailing piles, build a tiny lip at the bottom, and set up a rail?

Kiko and I hiked home. Kiko took a nap. I got on Google. Turns out, I wasn't the first person to try to build a rail setup in my backyard. In most of the videos I found, a seven-foot, thirty-five-degree in-run generated more than enough speed for a skier to slide a rail. As suspected, plastic grass was involved. But unlike the water ramps at Hood, these backyard versions were using the kind of artificial grass sold at Home Depot. Even better, in the videos, people were sliding on PVC pipes, which are cheap, light, and, when the inevitable crash came, definitely softer than steel.

Then I had a second thought—Mount Hood. Were they still running summer ski camps up there during COVID? The camps were typically for pro skiers or wannabe pro skiers. The average camper age was around seventeen. The coaches were maybe twenty-three. Was there a freestyle camp for fifty-year-olds?

I spent more time on Google. All those answers turned out to be "yes." There were freestyle camps held on Mount Hood. Most of the summer belonged to the kids, but there were two sessions aimed at adults. The June camp was COVID canceled, but the July camp was still up and running.

What if I could build dirt rails behind my house and start training now, then take my newfound skills up to Mount Hood for some midsummer polishing? Rail sliding, that definitely counts as progress. Never mind that it was a form of progress I'd never considered before, and a form of progress that most people considered impossible for a guy my age. If I learned a few tricks over the summer, that would definitely make me a better skier before next winter.

Not giving myself time to think about it, I reserved a spot at ski camp, taking note, as I punched in my credit card number, of its no-refund policy.

And this brings us to a Punk Rock Digression.

From the outside, the punk movement looked to be about sex, drugs, anger, and funny haircuts. But that's window dressing and not the window. The movement itself was about equality and creativity. Punks didn't care what you looked like, where you came from, or who you fucked.

What we cared about was DIY creativity—emphasis on *DIY.*

Punk was about doing it yourself. It was about turning anger into creativity and creativity into salvation. Punk was about starting a band before you knew how to play the guitar. It was about saying "screw you" to the music industry and birthing our own industry. Punks created their own record labels, put out their own records, booked their own tours, designed their own album covers, and published their own magazines to tell others about the scene. The first magazine I ever worked for was called *Penumbra,* which is a term that describes the lighter portion of a shadow, and a fact I know because I started that magazine. I was twenty-two years old at the time, and had to start my own magazine to get into the writing game. Back then, my dreadlocks hung to the middle of my back. Looking like that, no other magazine in America would give me a job.

My point: I learned to ski on converted garbage dumps. I was about to learn to rail-slide in an abandoned gold mine. If nothing else, that, at least, was punk as fuck.

## PINE NUT MOUNTAINS, APRIL 2, 2021

What if there was never a next season? In the middle of the lockdown, this was a real concern. What if the ski resorts never recovered; what if the chairlifts never ran again; what is the sound of one ski clapping?

No question about it, if I was going to be prepared for these possibilities, I was going to have to learn to enjoy hiking in the backcountry.

Yet, as mentioned, I hiked in the backcountry all the time, taking the dogs out on a nearly daily basis. These are always hard hikes at a fast pace. There's no choice. After much experimentation, I can now say conclusively: If the goal is tiring out a one-hundred-and-twenty-pound Maremma sheepdog, then three hill climbs with about two thousand feet in total elevation gain are the minimum requirement.

I loved these hikes. I loved hiking—just not when skiing.

First, hiking up snow is significantly harder than hiking on soil.

Not only is the terrain more tiring, you have to hike in ski boots, carrying all your gear, on slopes that are often damn steep. There is vertigo to contend with on the way up, but even if that was manageable, the real problem is the way down.

I sought progression, which meant pushing myself into harder terrain. Yet pushing my skiing when exhausted from hiking was an easy way to get hurt. If I wanted to push myself in the backcountry, I would have to attain a level of fitness that had long eluded me.

I have worked out three to four times a week since my freshman year in college. I lift weights for strength, do yoga for flexibility, and hike the dogs over uneven terrain for balance, agility, and endurance. There's no time for anything else. My schedule is always packed. My workday starts by 4:00 a.m. and ends after 5:00 p.m. In other words, more time for more training was a nonstarter.

But this is not a new problem, nor was it only my problem.

It turns out, when I'm not skiing or writing or hanging out with dogs, I'm also the executive director of the Flow Research Collective. We're a neuroscience-based peak-performance research and training organization. On the research side, we collaborate with scientists at places like USC and Stanford to try to figure out what's going on in the brain and the body when humans are performing at their very best. On the training side, we take what we learn from the science and use it to train everyone, from Fortune 500 companies to members of the US Special Forces community to the general public. But there's one commonality among most everyone we train—they're all busy.

Thus, at the Collective, we hunt *multi-tool solutions,* or solutions that solve more than one problem at a time. For example, meditation is a multi-tool solution. It trains focus, which helps produce flow, and it lowers stress levels, which often block flow. Two problems solved for the price of one.

Additionally, we hunt ways to *stack practices*, or nestle multiple training requirements into a single activity. Hiking my dogs uphill, at pace, over uneven ground, is an example. The dogs need to be hiked anyway, and harder hikes mean shorter trips to the gym. Slogging uphill fast trains endurance, coming downhill fast, over uneven ground, trains strength, balance, agility, and fast-twitch muscle response.

This last bit was key. Both strength and fast-twitch muscle response decline over time. This is one of the main reasons people think that learning to park ski is impossible for someone my age. But, like many stories in human performance, this one only tells part of the truth.

The actual truth is that strength and fast-twitch muscle response do decline as we age, but only if we're not actively training strength and fast-twitch muscle response. It's a use it or lose it situation.

This is not to say that these capabilities don't deteriorate over time. Muscle fibers begin to decline in number once we reach the age of fifty, but—if properly trained—those lost muscle fibers are buffered by the overdevelopment of ones remaining, or, as University of Michigan physiologist John Faulkner wrote in a 2008 meta-analysis of the issue: "Even with major losses in physical capacity and muscle mass, the performance of elite and master athletes is remarkable."

And today, standing on my porch and gazing at the mountains, that's when it hit me: I didn't have to add anything to my schedule. There was an additional practice I could stack on my current stack: weight vests. If I wore a weight vest while hiking the dogs, I'd train the same muscles needed to hike with a backpack in the backcountry without changing my schedule one iota. All that was required was mental fortitude: the willingness to let my regular hikes become harder slogs. In other words, I had to do what I was always telling other people to do if they wanted more flow in their lives: Get comfortable with being uncomfortable.

### NORTHERN NEVADA, APRIL 3, 2020

Two feet of fresh snow fell overnight. That I would not be skiing this snow broke my heart. Of course, unless you're a skier, by this point, you're probably wondering: What's with all the skiing?

Fair question. Now it's time to talk numbers.

There's a quiet mathematics to this life, a set of numbers that few of us want to discuss and even fewer bother to calculate. This isn't surprising—the math is ruthless. The numbers aren't the problem. It's the truth on the other side of those numbers.

For me, that math starts with skiing. In simple terms, skiing is my ultimate life hack. Why? Because skiing is the best I get to feel on this planet. In psychological terms, skiing represents my ultimate intrinsic motivator, the upper limit of my emotional possibility space.

How's that for a ruthless fact?

Sure, I tested other options. I tested—and this is a technical term—the *ever-living bejesus* out of these other options. I ran all the requisite sex, drugs, and rock and roll experiments. I've also weighed skiing against success, against love, against all the other things that I have been told should weigh more. . . .

The truth? No matter what I'm measuring against, skiing consistently emerges as the very best I get to feel on this planet.

This isn't about pleasure. If this was just about pleasure, well, her name was Lola, she was a showgirl. No, this is about meaning. When I'm skiing, when I push myself beyond the blurry edges of my abilities, life means more.

It just does.

Because the math is very precise, to truly dial in this meaning, I must ski under exacting conditions: in deep powder, at silly speeds, headphones blasting the Wu-Tang Clan, typically alone, typically in the trees, often in places where they'll never find my body if this shit goes wrong.

How's that for another ruthless fact?

Yet from a peak-performance perspective, there's peculiar power in this math. If we can figure out the thing that makes us feel the very best on this planet, we can use this information to steer. We can use it to help us solve life's ultimate puzzle—the fact that life is so short and, as Pablo Neruda says: the forgetting so long.

And I don't mean "solve" in a metaphorical sense. I mean solve in the only sense that matters: our yeses and our nos. Those things to which we say yes; those things to which we say no; the very algorithm that determines the quality of our days.

In my case, I say yes to skiing and I say no to anything that stands between me and skiing—which, I know, makes me sound psychotic. Let's build on this foundation. Yes, I also do all the psychotic stuff required to maximize skiing. I train like mad during the off-season. I also have to adhere to a serious recovery protocol—saunas, restorative yoga, Epsom salt baths—that allows me my madness.

All of this may seem a little over the top, but knowing I feel this way about skiing allows me to never waste time on less meaningful activities. Skiing is my first filter: If an opportunity presents itself, will it help me ski more frequently? Seriously, I really think this way.

Sure, I have a few other filters: writing, flow research, my family, my friends, making the world a better place for animals—but that's where this story ends. Alongside skiing, these are my six filters. They determine all of my yeses and all of my nos.

From a peak-performance standpoint, filters like this save us massive amounts of time. They allow us to avoid a considerable amount of decision fatigue, and truly align our lives with the three pillars of well-being: passion, purpose, and meaning. But more than any of that—there's the math.

Right now I'm just north of fifty-three years old. If actuarial tables are to be trusted, I have maybe thirty years left on this planet. Those are the basic numbers.

Now, I come from a long-lived family, take reasonably good care of myself, and have a number of scientist friends working on seriously whiz-bang longevity technologies, so maybe I'll get lucky and add another ten years to this particular ride.

So, let's set the number at forty more years on this planet, give or take. As it was 2021 at the time of this writing, that puts my end zone right around 2061.

Now, let's flesh out the numbers. If the best I get to feel on this planet is when I'm skiing through deep powder, well, really deep powder days only show up about seven times a season. And seven times forty—my years remaining—is two hundred and eighty.

That's it. Two hundred and eighty more chances to feel the very best I get to feel on this planet. Two hundred and eighty times to do the thing that makes life worth living. My point—two hundred and eighty is not a particularly big number.

Whenever I ski, I try to leave everything on the hill. I want to ski 'til I drop every time I get a chance to ski.

Why? Because life is short, and I did the math.

# Chapter 2

### NORTHERN NEVADA, APRIL 5, 2020

I didn't know if I could do hard hikes with a heavy weight vest—the jury was still out on that as well.

I messed up my back about a decade ago and it remained prone to the occasional meltdown. How much weight could I carry uphill without throwing it out again? That was still unknown. But if I was serious about skiing progress, I was about to discover the answer.

I decided to make a few rules. My plan was an inversion of most workout plans. I wasn't interested in the younger person's approach: fast progress and faster gains. Instead, my goals were slow, steady, and safe. Yet slow, steady, and safe would stretch out my learning curves, so I needed to do everything I could to preserve intrinsic motivation along the way.

To those ends, I loved my morning hikes with the dogs, especially the way my mind would drift and wander, solving difficult problems along the way. But my brain only drops into drift mode when calm and relaxed. If I was huffing and puffing, my muscles used too much energy for my mind to wander. This meant I could use mind wandering as a measure of fitness. Once I could do an entire hike with my mind in drift mode, it was time to increase the intensity of the hike.

Thus, the plan: Start with a ten-pound weight on my back and a half-hour hike, then work up in ten-minute intervals. Once I could hike uphill with ten pounds for thirty minutes with my mind wandering throughout, I'd increase the time by ten minutes. Once the hike got up to ninety minutes, I'd increase the weight by five pounds.

Then I set some fitness goals.

Normally, for the fitness levels required for fast geometry, I have to be able to ski a minimum of twenty laps in a day. Normally, it takes about twelve days to ski myself into the shape required to ski those laps. As fast geometry is my fastest path into flow and flow is the fastest path for progression, those twelve days are a hard limit on the quality of my ski season. But if I was fit enough to play fast geometry by day three, my flow total would increase by nine days, and that would definitely up my chances of making real progress.

Additionally, I wanted to be ready to ski on a daily basis. Even if the resorts managed to open their doors again, a COVID flareup could close them just as quickly. Plus, my record for days on the snow during a given ski season was forty-three and I desperately wanted to beat that record by at least a week. Fifty days on the snow was my minimum acceptable goal for the season.

Unfortunately, fifty days on the snow would be a stretch. That bout with Lyme disease robbed me of my recovery abilities. Ever since I got sick, while I can ski hard for a single day, I often need the next day off to recover. Occasionally, I get two days in a row, but performance declines precipitously by the second afternoon. Three days in a row is a feat I hadn't attempted in twenty years. Point of fact: To accumulate the forty-three days of my record season, I started skiing in early December and kept skiing through late May. For the season I was craving, I had to be ready to charge hard on a daily basis for fifty days.

"Have you considered prayer?" asked the voice in my head.

I had no idea what achieving this miracle level of fitness would take, but figured if I stuck to my normal workout routines and continued lifting weights or doing yoga three days a week; if I continued with my weight vest hikes until the weight was up to thirty pounds, and the hikes were ninety minutes long; if I could maintain this workout schedule, and those thirty-pound, ninety-minute hikes for five days straight, then I might be in the neighborhood.

Time to visit the neighborhood. I made my rules, set my goals, dropped a ten-pound weight into my backpack, and headed out the door. Turns out, a thirty-minute hike with a ten-pound weight was easy. The hard part was keeping myself in check. I got to the thirty-minute mark and thought: *That was nothing, not even a snack; let's keep going.*

Normally, I give in to the voice. In fact, I always give in to that voice. But slow, steady, and safe was my rule—so no snap decisions this time. This time it wasn't about my constant need to prove my manhood to myself. No, this time it was about my constant need to prove my manhood to the jocks who tormented me through my childhood, which is why I have the constant need to prove my manhood in adulthood. Wait, no, this time it's about . . .

As I said: Old. Dirty. Shame.

### PINE NUT MOUNTAINS, APRIL 7, 2020

Today was supposed to be an hour-long hike with a ten-pound weight, but my mind was so busy wandering that I got lost. In the end, I hiked for two hours. My legs felt fine afterward.

See, voice in my head? Miracles do happen.

### PINE NUT MOUNTAINS, APRIL 10, 2020

Fifteen pounds for fifty minutes and my legs felt solid—but having heavy weights loose in my backpack has been messing with my balance. Sooner or later, I was going to roll an ankle. It was time to upgrade my technology.

Then I got lucky. I was poking around the Google when I encountered the Aduro Sport. Unlike the standard full-torso weight vest, the Aduro Sport only covers the upper half of the chest and back, and just barely. Most weight vests are bulky contraptions meant to hold metal bricks. You can add or subtract bricks to increase or decrease weight. The Aduro is extremely low profile—just neoprene straps stuffed with iron filings. There's no way to add weight. If I wanted to increase or decrease the load I would have to buy another vest. But holding the weight on my upper back might protect my lower back, or so I hoped.

I started with the twenty-pound weight vest. Not bad. Pretty comfortable and my back agrees. Even better, because the vest sits on the upper torso, every time I took a step, I had to use all of my core muscles to stabilize the weight. It was like doing weighted leg raises, situps, and micro squats all at once—a perfectly stacked protocol.

I didn't know it yet, but in the season ahead, this perfectly stacked protocol—and the core muscles it strengthened—would save my life on multiple occasions.

## PINE NUT MOUNTAINS, APRIL 25, 2020

My love affair with the weight vests continued, but my rate of progress had not. With a twenty-pound weight vest, a twenty-minute hike worked just fine. Thirty minutes was dandy. But forty minutes had become some kind of psychic threshold. I'd been stuck here for over a week, with no progress in sight. Today, I was huffing and puffing the whole time. Instead of drift mode, my mind stayed in bitch mode.

Well, nah, nah, nah, not listening. . . .

**NORTHERN NEVADA, APRIL 27, 2020**

Lynsey Dyer called—it was fortuitous.

One of the best female big mountain skiers in history, Lynsey was a member of my peak-performance brain trust. The members were mostly women, and mostly professional athletes.

I prefer women to men in my peak-performance brain trust because women are often better than men at detecting the body's subtle emotional signals, or what's technically called "interoception." This skill translates into greater self-awareness and emotional regulation, both of which are crucial for peak performance. Plus, women don't tend to freak out when you ask them about their emotions.

Ask a male athlete how he dealt with the fear of skiing a big line, and he'll either deny the emotion or bro-brah the answer. Women can break these experiences apart. Ask Lynsey how she jumped Fat Bastard—a seventy-five-foot cliff in Jackson Hole, Wyoming, and a feat believed impossible for a woman to accomplish until she proved otherwise—and she'll talk about her physical, mental, *and* emotional preparation. She'll detail her off-season fitness program that got her legs strong enough to handle the impact of the jump, her off-season visualization program that got her brain ready for the huge rush of fear that would accompany seeing the cliff, and her off-season emotional fortitude program, which allowed her to push through that fear and fully commit to the line.

I told Lynsey I was contemplating learning to park ski. "I have a theory about how to do it safely," I said, as if that explained it. Lynsey had questions. I laid out my ideas, starting with the foundational fact: the sliding spin I'd thrown on Oops.

As a sliding spin involves nothing more than twirling 360 degrees with your skis on the snow, it's a relatively safe trick. There's no air time required. In fact, once you learn to keep the skis perfectly flat—to avoid catching an edge—it's hard to screw up. Skiers stop throwing them because of the cool factor or, rather, the lack of cool factor. Sliding spins are too easy to be steazy.

Yet spinning, even on snow, is an unusual sensation—novel, unpredictable, and mildly risky. All of these experiences push dopamine into the brain. Dopamine enhances focus, pattern recognition, and fast-twitch muscle response and primes us for flow. Why? Because focus, pattern recognition, fast-twitch muscle response, and flow are all abilities that help us survive novel, unpredictable, and risky situations.

Thus, while most skiers halt their sliding spin ways by their teenage years, dopamine was the reason I never stopped. I'd always twirl a few times at the front end of any ski session to prime myself for action. This was also why, when I saw that snow pillow on Oops, my brain suggested the sliding spin. The advice wasn't all that crazy; my brain simply saw a new place to throw an old trick and said as much.

I told this to Lynsey, then I told her my plan.

The sliding spin is my bedrock: a hardwired movement pattern that I can execute with little fear and a solid chance of success. If I build upon this foundation slowly, never asking myself to make more than one microscopic improvement to an existing motor pattern, then practice that new pattern until it's a hardwired code that can be executed with zero conscious interference, the results will be safe, incremental progress.

"It's like invading a foreign country," I said. "Except, instead of storming the castle, I'm sneaking in the back door, one inch at a time."

"This is your theory?" she asked.

It's not just my theory, I explained. It's what the science of movement has said for years. We don't see it used much because the progression ladder is too slow for most coaches. This system is designed to turn bad athletes into good athletes. These days, most coaches don't have this kind of time. Instead, they weed out the bad athletes and get them off the team.

But the limitations that make someone a "bad athlete" are the same limitations that make someone an "older athlete." Both bad athletes and older athletes have the same issues: less strength, limited reaction times, and difficulty onboarding novel motor patterns. As a result, bad/older athletes also have lower appetites for risk and increased fear when taking risks. In the face of physical challenge, all of these factors boost the production of stress chemicals like norepinephrine and cortisol. Unfortunately, these chemicals block learning, hinder performance, and make dropping into flow nearly impossible—which is why most action-sports athletes reach middle age and give up on progress and take up golf.

"What's step one?" asked Lynsey. "If the sliding spin is your foundation, what tricks are you going to learn first?"

I had that answer too. Henrick Harlaut gave it to me.

One of the greatest freestyle skiers in history, Henrick Harlaut helped pioneer a trick known as a "nose butter." To execute a nose butter, as you approach a jump, you lean over the noses of your skis, pressing them into the snow. As you fly off the jump, the noses snap straight, with the added force of the recoil propelling the skier even farther into the air. The trick is called a "nose butter" because the noses of the skis perform the same press-and-spread motion used when buttering bread.

In watching videos of Harlaut throw nose butter 360s, my body recognized the motion, and I recognized the recognition. But where was this recognition coming from?

Almost no other trick those SLVSH athletes were throwing made any sense. A cork 1080 with an octo grab—meaning, three sideways flips while holding the tail of your right ski with your right hand and the nose of your left ski with your left hand—in like what universe is that even possible?

But watching Harlaut throw a nose butter felt different. It felt familiar. And that familiarity was what got my attention.

After slow-motion video review, I figured out why. The first part of any of Harlaut's nose butters was a sliding spin. Harlaut would execute a sliding spin 180, lean over the noses of his skis, then launch off the lip and continue to spin. But I already knew how to throw a sliding spin 180. That move was on lock. So if I started practicing leaning over the noses of my skis during a spin—that was my next tiny chunk. My one-inch invasion of the world of park skiing.

"Nose butter 360," I said. "That's where I'm starting."

## NORTHERN NEVADA, MAY 4, 2020

Lynsey had asked an interesting question—what tricks did I want to learn? If I was going to learn to park ski, I needed to know more than a nose butter 360. I'd also have to learn to ski backwards and throw 180s, as those were the standard prerequisites for the trick. But what else had my attention?

And did I only want to progress my park skiing? After fourteen years in New Mexico working on tight trees, steep chutes, and other standard skiing challenges, was I going to abandon my big mountain ambitions? Not a chance.

Yet me deciding to ski a line during a given season was risky. It was like me deciding to throw a 360 during a given season. Or, for that matter, like me deciding to set any other kind of challenging physical goal—that is, historically problematic.

Just the act of setting ski goals could ruin my season. If I set a goal to ski a line, every time I saw that line, it would remind me of my still unmet goal. Fear would be the result. Over the course of the season, this could mess with my head and block my progress. Yet if I gave in to the fear and abandoned the goal, every time I looked at that line all I'd feel is shame. Old. Dirty. You know the drill.

There was no other choice. For peak performance, goal setting is critical to progress. Humans are goal-directed creatures, which is why research by psychologists Edwin Locke and Gary Lathan, the godfathers of goal-setting theory, shows that a properly set goal can boost motivation by as much as twenty-five percent. If I was going to learn to do something as difficult as park skiing, I was going to need that extra motivational fire. What's more, when battling fear, goals are the best weapon for the fight.

Our fears and our goals are the brain's primary filters on reality. As a result, these experiences are often binary. If I ski into a line with the goal of skiing fast, my brain filters reality to reflect this goal, specifically highlighting places on the mountain where I can find the most speed while producing the emotional experience known as "excited curiosity."

Additionally, as humans are also visually oriented, goal-directed creatures, we're hardwired to go where we look. When I stare at a spot with excited curiosity, my brain steers my body toward that spot, running an ancient two-step motor-action plan known to every toddler as, "Wow! Shiny thing! Must put in mouth for further investigation."

Using our visually oriented, goal-directed nature to our advantage is another secret to playing fast geometry. You don't steer the body with your muscles; you steer the muscles with your eyes—which is the reason you can steer at high speeds.

The opposite is also true. If I ski a line while fearing the line, my brain filters reality for all the scary stuff that I want to avoid—rocks, trees, the gnar—and my experience is one of growing anxiety. And since we're visually oriented, goal-directed creatures, when we stare at stuff that scares us, we steer directly toward that stuff. One moment you notice the gnar and get a tingle of anxiety, the next moment you're skiing the gnar, feeling terrified, and wondering how things got so bad so fast.

Well, now you know.

This meant, for my foray into Gnar Country, I would have to set serious ski goals to generate the motivation needed to achieve those goals. At the same time, all of those goals would have to build on preexisting motor plans in bite-size chunks—meaning, chunks small enough to keep fear at bay.

I'd have to choose my lines carefully. I couldn't overreach. Yet there were major zones at Kirkwood that I had yet to explore: the Palisades, the West Shore, Devil's Corral, Thunder Saddle, to name only a few. How not to overreach if I didn't know where I was going—that was the issue.

Additionally, there were a handful of revenge lines that I wanted to clean up. These were lines I skied badly last year, but never got to try again because COVID abridged my season. How to keep fear at bay if I'd already gotten my ass kicked by the line—that was another issue.

Finally, there were a handful of new-old lines or, essentially, all the places I'd already overreached. These were the lines that I had decided to attempt during the 2020 season, but my attempts had been thwarted by either my propensity for vertigo or my propensity for chicken shitting in my ski pants. How to try again when I'd already overreached and was now terrified of the line—that was yet another issue.

Most importantly, I could not waver. If I stopped working toward my goals and opened the door to fearing my goals, I'd ruin the season. The fear would creep into my subconscious, overpower my mood, and make finding flow nearly impossible. To combat this, I'd have to make actual progress every time I hit the hill. The kind of progress with which the voice in my head couldn't argue.

Done right? My motivation would skyrocket and my learning curve would creep along, and that was just fine with me. There might be the odd miracle, a trick I got the first time out. But if I could keep the fear away, while there would still be crashes, setbacks, and nagging injuries, I would definitely end the season as a much better skier than I was at the start.

Done wrong? If I started missing those goals, if I started fearing those goals, then that added fear would seriously hamper performance—which was an easy way to end my ski season in the cozy embrace of a morphine drip.

I made my trick list. I made my line list. The tricks all spoke to me aesthetically; the lines were all the grudges that needed settling. Both would increase my motivation and speed my progress.

Of course, the real goal was to learn tricks in the park, take them into the wild, and turn the entire mountain into a slopestyle course. But that was a long-term goal.

The short-term goal was simpler: I wanted one steazy line. More specifically, as terrain park lines come in sizes—kiddie, small, medium, large, and extralarge—my goal for the season was to try to learn enough tricks to throw a steazy trick off every feature on the first three of those options: kiddie, small, and medium.

If I could do this, then I would go from absolute beginner to low-level intermediate in a single season. This would be significant. I would be a radically different skier.

Was that too much to ask?

I began my list with eleven words that strike fear into the hearts of men, or, at least, into the heart of this man: "Learn every trick and ski every line on the list below."

## Trick List

learn to ski switch ("switch" is the technical term for skiing backwards)

180

snow slide 180 to forward and switch

nollie 180

ollie 180

nose butter 360
360
daffy butter
lazy boy hand drag 180 and/or 360
box slide to both forward and switch
dancer box surface swap 360
rail slide
shifty
safety grab
tail tap/tail drag
nose tap/nose drag

## Line List

Ski School Chute
Cliff Chute
Main Finger Chute
Eeny Chute
Miny Chute
Moe Chute
Upper Norm's Nose
Main Nostril
Side Nostril/left chutes
Side Nostril/right chutes
the Notch (from top)
Oops Cliff
Palisades
Devil's Corral
the zone between All the Way and the Notch
the zone between Pencil Chute and All the Way
Pencil Chute
Lightning Cliffs (jump the damn cliff)

I finished my list. I took a deep breath. I read my list. I rolled a joint. I smoked the joint. I took another deep breath. I reread my list. I rolled another joint. I smoked another joint. I took another deep breath. . . .

"Yo, big wuss," said the voice in my head. "You got this."

## PINE NUT MOUNTAINS, MAY 20, 2020

Ryan and I set up shop by the old gold mine in the Pine Nut Mountains. It was our first attempt at dirt skiing.

A twenty-foot-high tailing pile became our in-run. We cleared rocks, raked dirt, and laid thick cardboard boxes down the slope. We placed artificial grass atop those boxes and used metal stakes to anchor everything to the ground.

The tailing pile dropped down onto a sandy plateau, and that's where we built our jump. It was two feet high, made of dirt, covered with cardboard, and coated in plastic. Finally, we laid the PVC pipe a foot from the jump, anchoring it to the ground with more metal stakes.

Afterward, we surveyed the results. There was a single line of green plastic grass running across the desert. It looked like someone had built a solitary miniature golf hole in the most random location on Earth. It did not look skiable.

"Ghetto," said Ryan.

"Ghetto," I agreed.

After climbing the hill, we clicked into our skis and looked over the edge of the ramp.

"Ghetto One to Mission Control," said Ryan. "We have a problem."

The tailing pile we'd chosen as an in-run was a lot steeper than either of us had anticipated. Standing at the top, our skis jutted right off the cliff. To make this work, we'd have to hop into the air, land perfectly flat on the artificial grass and directly between the metal stakes, then not move a muscle as we *straight-lined* down the in-run and sailed off the jump.

The "straight line" was an issue.

As advertised, when skiers "straight-line," they literally ski in a straight line. There are straight-line chutes between huge rock walls, where the run narrows to the point that turning becomes impossible. In-runs to jumps can also require straight-lining. Sometimes you need the extra speed that comes from straight-lining to clear a gap; sometimes you need the extra speed for the extra air time required by an audacious trick; sometimes—like this time—there was no other choice.

A few feet to the right of our in-run was a cliff. As you can't turn on artificial grass, if either of us freaked out and tried to slow down during the straight line—which is often how the body responds to the feeling of rapid acceleration—we'd pitch off that cliff.

Ryan was our crash test dummy. Ryan is a gifted athlete and while he claims to feel fear, rarely does his countenance give it away. When I get scared, I twitch like a marionette. When Ryan gets scared, he gets still.

Ryan stayed still for a minute, adjusted his chin strap, lowered his goggles, nodded once, then jumped off the ledge. He nailed it, as gifted athletes do. He skied down the slope and slid onto the cardboard, smooth as silk. He didn't try to slide the rail. He was just testing the ramp and getting a feel for the ride.

Now it was my turn. I stood at the top thinking, *You can't drop two hundred dollars at Home Depot, spend two hours digging in the hot sun, and not try this shit, right? I mean, there's a code.*

Run one went fine. And by fine, I don't mean elegant. Ryan made it look easy. I did not. But practice makes perfect, and I was down for another attempt.

Ryan led round two. This time, he skied down the ramp, hit the jump, hopped sideways onto the rail, slid sideways across the rail, then hopped switch off the rail—that is, he landed backwards—before gliding to a halt on the cardboard.

"Unbelievable," said Ryan. "It actually worked."

I stood at the top for my second attempt, thinking that sooner or later I was going to have to slide a rail, thinking of my trick list, my line list, and my sworn commitment to progress, thinking—enough with the thinking. I jumped off the ledge, slid down the ramp, popped off the lip, landed sideways on the rail, and started a trend that would continue for most of next year: I smashed, at high speed, into solid ground.

"Body armor," said the voice in my head. "That is why you brought body armor. Now put it on."

I donned the armor, but smacking dirt is still smacking dirt, and that was all I did all afternoon. Twelve attempts, eleven failures, one bit of incremental progress: I managed, on one attempt, to 50–50 the rail. In other words, I jumped onto the rail with my skis parallel, then skied straight across the rail. It's called a 50–50 because rails are skinny, and rarely do skiers have more than fifty percent of either ski on the steel.

While 50–50ing a rail can be tricky, on a wider feature, like a box, it's pretty straightforward—a fact I can share with confidence because 50–50ing a box was the only park trick I actually knew how to do.

And this brings us to: A Brief History of Steven's Terrain Park Career

Prior to today, I'd skied in terrain parks on only four occasions. The first time was in Whistler, in the mid-1990s, and long before anyone had figured out how to build terrain parks. All I remember was a one-hundred-and-twenty-foot tabletop jump that absolutely no one would hit.

The second time was in Squaw Valley—which is today known as Palisades Tahoe*—and also back in the '90s. I don't recall if I actually

---

* At the end of the 2021 season, Squaw Valley and nearby sister resort, Alpine Meadows, would be linked together by a gondola and renamed Palisades Tahoe. As everything in this book takes place before this occurred, and to avoid the confusion of having one name for two locations, I have kept the old names. Also Palisades Tahoe is a dumb name. It sounds like a gated subdivision in Orange County.

skied through the park, but—as snowboard culture was then obsessed with gangster rap—I do remember riding up the chairlift and seeing a snowboarder backflip a huge jump and watching a handgun sail out of his waistband. The snowboarder landed down the hill. The handgun landed on the top of the jump—and bounced.

My next terrain park foray was delayed by two decades. I spent the interim trying to improve my skills in the steep and deep and didn't venture into another park until visiting New Mexico's Red River Ski Area, in the crap winter of 2017. Snow levels were low, the hill was frozen solid, and there was nothing to do but explore the park. I recall a sordid attempt at the big line, a sorrowful attempt at the medium line, and then, in the small park, on my fourth try, I managed to clear all three of their tiny tabletops.

My final terrain park visit was at the Santa Fe Ski Area, where they consistently build one of the worst terrain parks imaginable. Jumps were forbidden. Boxes—that is, long, skinny wooden rectangles coated in slippery plastic—were okay. Thus, in 2019, on the last day of my last season at the Santa Fe Ski Area, I accomplished a feat most park skiers pull off before their thirteenth birthday: I 50–50'd my first box.

And today, here in the fading glory of an abandoned gold mine, in the late spring of 2020, at the tender age of fifty-three, I managed to 50–50 my first rail. I took an existing motor pattern—my ability to keep my skis together while moving in a straight line—and transferred it to a novel location: the dirt rail we'd set up.

"Unbelievable," I said afterward. "That shit actually worked."

"Ghetto One to Mission Control," replied Ryan. "We have liftoff."

## NORTHERN NEVADA, JUNE 1, 2020

COVID closed the gyms in March, but they reopened again in June. As I was on a mission, I decided my workouts were worth the risk. But I changed my approach.

My old workout was three times a week, three sets of ten reps, fifteen sets total, hitting shoulders, back, legs, chest, and abs in a single go. My new schedule was four days a week and focused on split training. I trained my core and my legs every session, but then split back and shoulders one day, chest and arms the next, five sets of ten exercises per body part.

Also, I wanted hard data, so I took some baseline metrics. Weight: one hundred and forty-four pounds. Max bench: two hundred pounds, five reps. Max leg press: three hundred and fifty pounds, ten reps. I wrote everything down in a notebook. Were these big numbers?

For the ski season I had planned—not big enough.

## PINE NUT MOUNTAINS, JUNE 5, 2020

A ninety-minute hike with the twenty-pound weight vest was the goal. The voice in my head had a lot to say about the idea. For the first ten minutes of the hike, I got the play-by-play: "Your back hurts, your legs hurt, your abs are tired, it's too hot, it's too early, and do you remember how the jocks used to punch you in the arm during math class—you fucking wimp, you just sat there and took it."

And repeat.

Yet once you get beyond the aesthetics, getting old has its benefits. Over time, we accumulate decades of practice ignoring the voice in our head.

I finished the hike: twenty pounds, ninety minutes, not a problem. Then I repeated the feat every day for a week to make sure it stuck.

## NORTHERN NEVADA, JUNE 9, 2020

I got a text from an old friend who lived near Squaw Valley. It triggered a flood of memories, aka Steven's History with Ski Towns.

In the 1990s, right after college, I moved to Aspen, Colorado. I didn't know much about Aspen. I'd moved there for only one reason: There was an exceptionally well-stocked bookstore in town.

Back then, I had two desires: finish my first novel and get better at skiing. The novel required research and this was pre-internet, when most ski towns were podunk places without decent libraries. But that Aspen bookstore was huge, with a coffee shop attached. I figured I could use the bookstore as a place to do research, use the coffeeshop as a place to write, and use the mountain as a place to ski—then all would be right in my world.

All was not right in my world. Aspen was a town filled with trustafarians with coke problems and Hollywood starlet transplants with coke problems and snotty New York stockbrokers with coke problems and you get the picture and yes, the picture had coke problems. In my quest for a good bookstore, I'd ignored the fact that Aspen was, well, Aspen—which is why I split for San Francisco after only one season.

My next attempt to live in a ski town came a few years later. I was still residing in San Francisco but had finally earned enough money to afford a beater truck with four-wheel drive. Did the truck have enough stamina for the four-hour trek north to Squaw Valley, that legendary mecca of hot dog skiing? I was going to find out.

Affording lift tickets was my next problem, but then fortune smiled upon me. I figured out how to get free lift tickets. And free ski gear. And this bring us to *Freeze* magazine.

Until the early 1990s, ski media was dominated by stalwart publications like *Ski* and *Skiing*. These magazines served a specific function: lure wealthy East Coast vacation dollars to Western ski town cash registers. Mostly, they contained articles like: "The Five Best Intermediate Runs at Vail" or "The Five Best Places to Get Pizza in Park City." *Powder* magazine was also around, but I was an angry punk rocker and their soulful, hippie vibe wasn't my thing. And I wasn't alone—a fact that partially explains *Freeze* magazine.

Then again, nothing explains *Freeze* magazine, though this may take some explaining.

Before 1984, it wasn't just the ski magazines that were shackled by culture. Media belonged to the big three networks, a half-dozen newspapers, and a well-sanitized array of two dozen magazines. Cable TV, outside of a few major markets, did not exist. If you were a weirdo trapped in Tulsa, Oklahoma, or Fargo, North Dakota, or Cleveland, Ohio—well, the punk rock zines that began trickling out in the late '70s and early '80s were your only lifeline to the outside world, and sanity.

All that changed in the early 1980s, when four unrelated trends collided. First, punk splintered. When the term first emerged in the late 1970s, it was a derogatory catch-all for an array of musical styles. Sure, there were traditional "punk" bands like the Ramones, but in those early years, Blondie—or what today might be considered Swedish disco—was considered a punk band. So were the Talking Heads. The band Television is often considered the first punk band, yet today sounds like the Allman Brothers in a bad mood. Still, despite this sonic diversity, the mainstream populace hated punk as a category—so nobody bought albums by any of these bands.

Then some record exec had a brilliant idea: rebrand the softer, arty side of punk as "new wave." Maybe then, the thinking went, the kids will buy the product.

It worked. The product started to move. Flock of Seagulls, Culture Club, Wham!, Billy Idol, Thompson Twins, the Smiths, the Cure, New Order—all of these bands started gaining fans and making bank. If only there was a way for big advertising dollars to reach this new breed of new wave consumers.

Technology came to the rescue. In 1984, Apple introduced MacPublisher, the world's first desktop publishing program. Overnight, those people lucky enough to have a computer could do color graphics, arrange text in columns, and, miracle of miracles, spellcheck. The first people through this new publishing door were the people already through that door—the punks who had been putting out zines for years.

The punks had all that DIY experience, so they were the first people to colonize this new territory. Within five years, if you were countercultural and had access to money, starting your own magazine was the thing to do. By the early 1990s, on the eve of Nirvana, when the entire world was about to smell like teen spirit, in an effort to reach this strange, new wave market, corporations started pouring advertising dollars into these new magazines. Which is how, for a short period of time that started in the early 1990s and ended with the dot-com crash of the early 2000s, punk rock weirdos took over the media world.

Finally, the "extreme sports" movement, which was then a small subculture dominated by working-class, often California-based punk rockers, caught the mainstream's attention. The X Games got going, the Gravity Games as well. For the first time ever, skiers who *weren't* racers could get paid to ski. All that was required was a willingness to huck big cliffs in fall-and-die zones for the camera. Once again, punks, who had nothing to lose and few other prospects, were the first ones through this door.

And it was the confluence of all of these trends that birthed *Freeze* magazine, both the world's first extreme ski magazine and the first magazine to give me a title on its masthead—Squaw Valley correspondent.

Squaw was the center of this new universe. The terrain was ferocious and photogenic. The skiers were ferocious and photogenic. "Squallywood" wasn't for everyone, but for a certain breed of DIY creative with punk rock leanings and an adrenaline-junkie personality, it was home.

At the center of Squaw's allure was KT-22, aka "the mothership," aka the most famous chairlift in skiing. Enormous lines run everywhere off the top. Huge cliffs sit directly under the lift. On a good day, a skier can get twenty-five laps down KT, which is a level of practice in expert terrain that isn't achievable in many places, and helps explain why Squaw Valley consistently produces expert skiers.

Chute 75 is the most famous run on KT-22. Rated the "Fifth Steepest Run in Tahoe" by *The Liftie Report*, Chute 75 requires a cornice huck onto a forty-five-degree slope that funnels into a narrow choke before becoming one of the longer, rowdier bump runs in skiing. Though I left Squaw in 1997, I tried to return every year, in late May, so I could spend my birthday skiing slush bumps on Chute 75.

In fact, the COVID-halted season of 2020 was the first year I'd missed KT in a long while. This was why, while sitting at my desk and pining for the ski resorts to reopen, my friend's text caught my attention. He wanted to tell me about a secret mission. He'd been hiking up the backside of KT, sneaking into Squaw Valley and poaching lines down Chute 75—which, yes, still has snow in June—and did I want to join him for this somewhat illegal adventure?

I was desperate to ski, but the idea of hiking up the backside of KT scared me. I didn't want to arrive at the top and be too tired to ski down again. But if my weight vest workouts were actually working—wouldn't this serve as the perfect test for my fitness?

Plus, if I wasn't tough enough to hike up KT, then I wasn't tough enough to learn to park ski, then I wasn't tough enough to prove my worth to all those people who doubted my worth. . . .

"I'm in," I texted back. "I'm bringing dogs along for moral support."

## NORTHERN NEVADA, JUNE 10, 2020

It was late afternoon when we started our hike. Two dogs, two dudes, and a long climb to the top. Actually, it wasn't that long. There were a couple of steep sections early on, but the weight vest hikes had gotten me used to the acid burn of the uphill. Once my brain realized that my body was strong enough to handle the effort, my mind chilled out and I could check out the scenery.

Craggy, snow-coated peaks and ancient pine trees—that is, majestic, in every direction.

Then we rounded a bend and my friend—who will remain nameless as what's about to happen is technically illegal—suggested a shortcut. If we scrambled up a couple hundred feet of boulders, we'd trim forty minutes off our hike. The boulders weren't all that steep, but they were seriously exposed, and that worried me. Empty space in all directions, especially in unfamiliar territory, was a recipe for vertigo.

It didn't take long.

I was halfway up the first boulder when my vision started to cloud, and the world started to spin. Those ancient pine trees became whirling dervishes. I felt my pulse rate spike. I started to lose muscle control. I stood very still, trying to catch my breath and calm my nerves. When that didn't work, I started swearing.

"Motherfucker!" I shouted.

"You've got this," my friend called back.

"Pussy," said the voice in my head.

"That term is offensive to some people," I replied.

"What term?" asked my friend.

Long story short: The vertigo passed, the shortcut saved us time, and we made the summit in about ninety minutes. But the summit wasn't the goal. I hadn't been training for the summit. I'd been training to hike uphill fast and ski downhill faster and feel strong throughout.

We crested the summit and started hiking toward the top of Chute 75, only to discover the snow had melted out of the chute's entranceway. Rather than scrambling over bare rock to get to skiable terrain, we left the main entranceway and traversed over to the "First

Alternate"—meaning, the first alternate entrance to Chute 75 if the main entrance is impassable. As the First Alternate sits in the shade, there was still snow on the slope, top to bottom.

We clicked into our gear. My friend dropped in first. He's normally an incredibly powerful skier, but as I watched him descend, even though his turns were smooth, they were much slower than normal.

"Peculiar," said the voice in my head.

Before I had time to ask what was up, my friend was already two-thirds of the way down and out of earshot.

Now it was my turn. I pushed off, carved across the face, and discovered the problem. What looked like a two-inch layer of slush was actually eight inches deep. My turn unleashed a river of muck. Balls of wet snow and fist-size rocks of granite were bounding down the chute beside me. Ever try to play dodgeball in a swimming pool filled with mud? As the dogs were also chasing me down the hill—and coming dangerously close to the razor blades strapped to my feet—I now understood why my friend had skied so slowly.

But I didn't ski badly, and I wasn't too tired afterward. In fact, on the hike out, I felt pretty good. If I got to Mount Hood in this kind of shape, I'd definitely be ready to ski. If I also got to Mount Hood able to slide rails?

Now that would be real progress.

### PINE NUT MOUNTAINS, JUNE 20, 2020

Dirt Rails at the Gold Mine: Round Two, aka the Gold Mine Strikes Back.

To improve our terrain park design, we chose a mellower tailing pile. The lower slope angle made the in-run to our jump a lot less terrifying. We also found a way to anchor the plastic grass to the ground that didn't involve putting metal stakes directly into our flight path. This made the takeoff a lot less dangerous.

So much for our new design.

My first attempt ended with my next crash. The ground felt even harder than before. Then, on my second try, a miracle occurred: I slid my very first rail. It was ugly. It was wobbly. But I had done something on skis that I couldn't do at the end of last season, and that had been my initial goal.

Now all I had to do was dial in the trick.

Not so fast. I hit the ground twice in a row. On my fifth attempt, I tried to spin in my unnatural direction; that is, in the direction my body doesn't normally want to spin. That didn't work either—unless, of course, my goal was to land on rocks and start bleeding.

There was more bleeding on my next two attempts, both in my unnatural direction. I went back to my natural direction for two more tries. The result: more bleeding. Then, on attempt number eleven, I threw what ski scholars refer to as an "inadvertent backslide," meaning, while keeping my back ski on the rail, I lifted my front ski into the air—which would be damn steazy if I could learn to do it on purpose.

But not today.

Today my inadvertent backslide and my wobbly rail slide were as close to victory as I would get. Ryan nailed every trick he tried. I nailed the ground, again and again.

Everything hurt, especially my pride.

### MOUNT HOOD, JULY 3, 2020

Ryan and I drove to Mount Hood in separate cars to observe COVID protocols. We made great time to Bend, then caught traffic the rest of the way. The trip should have taken nine hours. It took thirteen.

Upon arrival, we checked in with the ski camp. Immediately, I realized my mistake. I'm an introvert. Strangers make me uncomfortable. Crowds of strangers make me seriously uncomfortable.

The camp was crowded. I was nervous. I could deal with the anxiety for the first few days, when Ryan was there for moral support, but he had to get back to work by Tuesday, and I was staying through Saturday. Once Ryan split, it would be just me and the strangers.

Let's just say the "welcome to camp" meeting did not make me feel welcome. It made me feel weird and alone, the way crowds always do. The camp was just too social an atmosphere for someone as antisocial as me.

Still, there was snow in the mountains, and I hadn't had a vacation in nine years. . . .

### MOUNT HOOD, JULY 4, 2020

We got on the mountain early, finding jaw-dropping nature in every direction. Mount Hood is a still-active, cone-shaped volcano with a cliff-lined, gates-of-Mordor kind of crest and a sweeping glacier below. As that crest is usually hooded with snow, hence the name.

Seeing Mount Hood again produced instant time travel. The last time I skied here, I was twenty-six or twenty-seven, or one of those other impossible ages that seemed so old at the time yet feel so young in retrospect. I was with the original bad-boy skier, Glen Plake, with his towering mohawk, gargantuan laugh, and incredible skills. In the iconic ski film *Blizzard of Ahhhs,* Plake's descent of a burly Chamonix Couloir—a forty-foot straight line down a fifty-degree slope into a forty-foot cliff huck that required a midair turn to the left to avoid rock splatting, and a you'd-better-stomp-your-landing-then-shut-'er-down-in-a-hurry amount of precision, as a nasty crevice stretched across the bottom of the run—changed history.

Plake nailed the line, stomped the landing, cartwheeled on impact, and ended up throwing an accidental back handspring over the crevice—and this is another difference between tourists and locals.

Tourists read this story as a cautionary tale about why not to jump off cliffs. Locals understand this saga as the beginning of the freeskiing era.

The thing about this story—there's more to this story.

Before Plake had his go, Scott Schmidt made an attempt. Schmidt was a laidback jock, an ex-racer with deep ability and a preternatural calm. Schmidt skied the line, stomped the landing, and also exploded. He too bounced over the crevice. It was such an insane feat that the film crew began to pack up their gear to leave the scene. But then Plake clicked into his skis, slid atop the couloir, and looked ready to give it a go.

The cameras were turned back on.

Plake actually took a rowdier line than Schmidt did off the cliff, adding an extra ten feet of air time to his trip. What was the big deal? Plake had outdone Schmidt, the widely acknowledged best "extreme" skier on the planet. Yet Plake wasn't a jock. He was one of us: a punk rock fuckup from Tahoe. Suddenly, the upper echelon of skiing didn't solely belong to the ski racers. It belonged to anyone with the skills and attitude to storm the castle. And yeah, in the punk scene, there was plenty of that going around.

Back in 2020, Ryan and I found Mount Hood's lift line jammed with little racer kids with lots of racer attitude. One of them skied over the backs of my skis and knocked me on my butt. I almost got mad, but then I remembered I used to be just like them, a "mini-shred" in the lingo. Back when I was a mini-shred, terrorizing adults helped make me into the adult I am today.

Hard to get mad at that.

I got mad at Hood's terrain park, which was the size of a postage stamp. Everything all pressed together in a single corner of the glacier. The line for the J-bar was long. The run was short. And I was nervous. There were too many people watching, and what was I thinking anyway? Park jumps, with their big lips and blind takeoffs, rattled me. Sliding rails, especially after getting my ass kicked by the dirt rails, raised that rattle to a quake.

My bout of bad nerves cost me. I barely cleared the first jump on their kiddie line and could never make up the lost speed and that was my first trip through the park.

We tried again, but round two wasn't much better. After a third trip, it was clear this wasn't working. Beyond my Chute 75 adventure, I hadn't skied in months. Ryan felt the same. We were both having out-of-body experiences, and not the good kind.

We bailed on the park and headed to the top of the glacier. A few fast laps under the lift to get used to skiing at speed, then we went exploring. We were hunting dopamine. Out past the slopes where the racers train, we discovered a huge quarterpipe that led to a line of machine-sculpted zipper bumps used by the pro-mogul camps.

This was the solution we'd been seeking.

Ryan is an air-hunting ski missile, meaning, on his way through the moguls, he found around five great jumps. We hit all five at speed. By the end of our second trip through the bumps, I started to feel a feeling I hadn't felt since the resorts closed. I was skiing. I was still among the living. It was a miracle.

After two more laps, my hips were loose, my stance was low, and my bounce was back. I felt jiggy. Put me in, Coach, I'm ready to rock.

Ryan agreed.

We skied toward the upper terrain park—an array of smaller features meant to serve as a warm-up for the larger fare to come. Maybe seven hits in total: a tiny right hip jump into a medium-size wall ride into a six-foot tabletop jump into two dance floor boxes and a friendly beginner rail.

Ryan blazed toward the first hip. I assumed he would "straight air" the jump to gather speed for the features below. Ryan had a different idea. He hit the hip and spun left, twirling an elegant 270 and glancing my way upon landing. I knew that look. Ryan was wondering if I would try the trick.

It was a good question.

I'd never thrown a 270 before. In fact, even though I've been desperate to learn to throw 360s for thirty years, I'd never even attempted a 270, the widely acknowledged first step in 360 progression. That's the thing about fear—it can steal decades from your life if you're not paying attention.

But I didn't drive for thirteen hours to sit on the sidelines. Plus, there was no way I was gonna get my old dirty shame on my shiny new plan. Instead, I just did with my body what Ryan did with his body, and my body did a 270. It was the beginning of a realization that would shape my season: If Ryan can do it, I can do it, and therefore I should.

This idea was not based on bravado.

Ryan and I are both expert skiers with similar body types, ski styles, and terrain preferences. Additionally, we both have injury-riddled pasts that give us considerable incentive for avoiding the hospital in the future.

I'm slender and have a fondness for tight trees—meaning, I like to blaze through small openings between hard objects. But if I leap between two trees and land between two more, Ryan knows—because he has a similar body type, and his body responds to terrain variations in the same way that I respond to terrain variations—that he can do it as well. And vice versa.

I also created three exceptions to my new rule, all meant to save me from myself.

First, if I was feeling too much fear—if the emotion was interfering with my performance—back off and come back later.

Second, pay attention to exhaustion levels. Once I start underjumping hits and making weak-ass turns, once again, I was done playing hero for the day.

Third, Ryan is a former sponsored athlete with a park skiing background. While we're evenly matched on the mountain, in the park, he can do stuff I can't—like sending 540s to the moon off extralarge kickers—so, for now, lunar explorations were strictly off limits.

But if Ryan made a move, and the move was within my reach, it became mandatory. This was the second realization that shaped my season: Always follow the rules.

I had goals. I had rules for achieving those goals. And the part that made the biggest difference: I worked for the boss.

Who's the boss? The boss is the version of myself who creates the goals and sets the rules. In the moment, especially if the moment is difficult, the version of me making decisions often wants the easy way out. The quick fix. The biggest high. But the boss prioritizes my long-term interests over my short-term pleasures. Learning to park ski was my long-term interest. Thus, if Ryan can do it, I can do it.

Ryan saw my 270 and slowed down to offer a high-five. I was about to slow down as well, but then I saw the next feature—the wall ride—and my brain had a different suggestion.

To understand this suggestion requires a brief introduction.

Meet James Jerome Gibson, a twentieth-century psychologist who spent his career studying visual perception. Gibson discovered that when animals look at the world, they don't see objects, they see *affordances,* or what the object offers the animal.

If a frog and a fish look at a lily pad, the frog sees a place to sit on, the fish sees a place to hide beneath—these "options for interaction with an object" are affordances. When I play fast geometry, noticing how a particular terrain feature will launch my body in a specific direction, I'm playing a game of affordances. More importantly, affordances aren't static; they develop as we develop, which brings us back to the wall ride.

A wall ride is exactly as it sounds: a big plastic wall jutting up from the snow. This particular wall was ten feet high and twenty feet long.

The easy option is to jump off the far edge of the wall, essentially using the feature like a slightly more vertical hip jump.

That had been my plan too, and it was an ambitious one, as this was the first wall I'd ever attempted to ride. But I was all fired up after my 270. More importantly, that Oops sliding spin 360 had altered my affordances. Thus, when I saw the wall, my brain recognized a pattern and suddenly it wanted to play fast geometry. The wall looked a lot like the wall of the Oops halfpipe—the spot where I'd attempted that sliding spin 360—so why not build on that attempt?

My brain suggested a sliding spin 540—ski forward into the wall, spin a 180 and ride backwards up the wall, then execute another 360 on the trip back down. But the rules for fast geometry never change: compute, then execute.

It worked perfectly. I spun a 180, slid onto the wall backwards, then spun a 360 on my way back down. Every motion was familiar. It was maybe the steaziest thing I'd ever done on skis. More importantly, it was proof of concept. Adding a tiny new movement to a long-hardwired motor action plan was the fastest and safest way forward.

As I skied away from the wall, Ryan cheered. Turns out, one inch at a time might actually get me where I wanted to go.

### MOUNT HOOD, JULY 5, 2020

Yesterday, the combination of my initial terrain park terror and my eventual wall ride triumph sapped all my strength. When I got back to the hotel, I turned off my phone and collapsed into bed.

Today, I woke to a dozen messages from my wife, each one more panicked than the last. The largest wildfire in Nevada history had ignited in the Pine Nut Mountains, less than five miles from our new home. At midnight, while I was dead asleep, my wife had been evacuated from our new home. Two of our dogs had gone missing during the evacuation. They were presumed dead. Our new home, presumed torched.

A final text message included a photo she'd shot midevacuation, taken from the window of a speeding car. It showed the sky on fire, a curtain of flame a half mile long and a hundred feet high.

I got in my truck and started driving.

# Chapter 3

## NORTHERN NEVADA, JULY 13, 2020

Our new home didn't burn to the ground. The wildfire was roaring down our street when a change in wind direction sent it back toward the mountains. Still, the danger wasn't over, and the fire department wouldn't let us back into our neighborhood for a few days. Once they lifted the evacuation order, we found our missing dogs, but the blaze wasn't under control, so we spent the next week on a sleepless fire watch.

While under fire-imposed house arrest, high-intensity workouts were the only way to stay sane. As soon as it was safe to be in the backcountry, I donned my weight vest and went to survey the damage. I could see the fire's path as I hiked: huge swatches of pristine wilderness now charred and black.

The long drive home from Mount Hood, my fear for my dogs, the terror of the fire watch, those long, sleepless nights—they all compounded matters. My park skiing efforts weren't going according to plan. The dirt rails produced more pain than progress. The ski camp was a bust. And my first vacation in nine years had just been spoiled by a natural disaster.

After teetering on the edge of despair for a couple of days, I surprised myself and went in a different direction. Instead of wallowing, I got online, ordered a thirty-pound weight vest from the Aduro website, and decided to double down on my training.

Was this grit? Was this resilience? Or was this the beginning of an unhealthy obsession?

And, seriously, how can you tell them apart?

### PINE NUT MOUNTAINS, JULY 18, 2020

The first of the thirty-pound weight vest hikes—thirty minutes over easy terrain just to get a feel for the added weight. I got a feel, all right.

Heavy. Very fucking heavy.

### STATELINE, NEVADA, JULY 25, 2020

His real name is Buck Brown, but they call him "the Wizard." Buck's a master boot fitter, a rebel, an artist, inventor, entrepreneur, and, technically, the owner of Olympic Boot Works, with locations in Olympic Valley, California, and Stateline, Nevada.

To understand Buck's craft is to understand the torture of ill-fitting ski boots. It's a torment that turns mature adults into blubbering messes, robs expert skiers of their most basic abilities, and continues to confound the entire ski industry, as they remained incapable of creating a technology capable of taking all the power generated by the leg and transferring it, without too much loss of energy or the feeling of having an appendage rammed inside a meat grinder, to the ski itself.

Buck believes in "the primacy of the heel." His is a contrarian's stance. There's a general philosophy in podiatry that the foot's arch is the source of its power. Almost all orthotics, including most custom-fitted ski boot liners, have emerged from this philosophy.

Not so fast, says Buck Brown.

Buck's philosophy starts with the heel. He believes that a great ski boot should lock the heel into place, thereby increasing its natural stability, enhancing its basic movement patterns, and aligning it with the body's kinetic chain. But what the hell do I know? I just asked around. Who's the best boot fitter in Tahoe? Everybody said the same thing: Go see the Wizard.

So, of course, I was off to see the Wizard.

The Wizard custom-fit me for one pair of boot liners, two power straps, two pairs of Technica ski boots—the Mach 1 for the alpine, the Zero G for the backcountry—and three pairs of socks.

Walking back to my truck in the parking lot, I felt a twinge of something I hadn't felt since returning from Mount Hood—hope.

I had needed new ski boots for years. Now I had them. I might survive, after all.

## NORTHERN NEVADA, AUGUST 8, 2020

The purchase of ski boots made me feel better. It made the ski season feel one step closer. More importantly, it made August—always my least favorite month: hot, dry, snowless, hopeless, endless, and made so much worse this year by the wildfires raging across the Western states—a little more tolerable. Thus, starting today and about once a week for the next three months, it was retail therapy to the rescue.

First, Armada sold me a new ski jacket. Next, Armada sold me a new pair of heavy ski pants. Finally, Flylow sold me a new pair of light ski pants. Then I remembered that, up at ski camp, the coaches raved about the freestyle performance of the new CT 2.0, a ski made by the Faction Collective.

I didn't have a pair of freestyle skis, and then I did.

Of course, I had to buy a pair of Look Pivot 15 bindings to go along with the new skis. And poles. And what about powder skis? I had my Pescados, but those were "superfats," meant for trips to British

Columbia or Japan or for those once-a-year days, with three feet of light, fresh snow. Would they work well with a foot of thicker stuff? How about in the chop and the crud? What about with Sierra cement?

Maybe a pair of Faction's Prodigy 4.0? I mean, they are all-black and match my new Armada jacket. Plus, the cool custom graphics on the ski—the shadowy outline of a dog. I mean, it's a Ghost Dog—you get the symbolism, right?

We'll come back to the symbolism. For now, let's set the scene: The entire western United States was on fire, plus COVID, plus a chaotic presidential election, plus no skiing, plus work stress, plus other stuff. Also, my attempts to smoke all the weed in Nevada weren't helping matters.

Maybe it was time for hard liquor? Or harder drugs? Turns out, retail therapy was exactly what I needed.

Those Prodigy 4.0s—yeah, baby, they were mine.

### PINE NUT MOUNTAINS, AUGUST 13, 2020

Dirt Rails in the Gold Mine, Round Three, aka the Ryan Wickes Show.

On our third attempt at rail sliding, I hit the ground hard, like eight times in a row. Along the way, I felt too many emotions: annoyed, irked, frustrated, furious, depressed, despondent, then, finally, in too much pain to care. I'll spare you the narration. Let's just say, the voice in my head had plenty to say.

Meanwhile, Ryan nailed ten rail slides going forward, then decided to try one backwards. One practice run to get used to skiing backwards down the ramp, one switch rail slide across an old PVC pipe at an abandoned gold mine in the soft twilight of a fading August afternoon.

This brings up another difference between tourists and locals. Ryan is twenty years younger than me. He was a star athlete in high school, an all-state quarterback who chose not to play sports in college, but as a skier, in the years afterward, made it onto the Tomahawk Apparel freestyle team. Tourists think the gap between being able to ski a run marked "experts only" and being an actual expert is actually a gap. Locals know it's an abyss. The reality is that I got one successful, forward-facing rail slide out of thirty-five attempts. Ryan went forty out of forty while going forward, then got one going backwards on his second try.

I tried not to think about our age difference too much. I tried not to think about our talent difference too much. What did Nietzsche say? "When you gaze long into an abyss, the abyss also gazes into you."

Yeah, he said it right.

## NORTHERN NEVADA, AUGUST 20, 2020

Of course, for the Prodigy 4.0s, I needed new bindings. Look Pivot 15s, here I come. Extrawide brakes please, so they'll fit around my new fat skis.

Of course, since I live in the middle of nowhere, we don't get our mail delivered at home. Instead, we use a P.O. box. Normally, my wife swings into town to grab the mail every few days, on her way back from the dog park. Today, I realized I needed to start running secret "package intercept" missions.

Either that, or I'd have to explain why I'd been draining our retirement account to finance my ski addiction.

## NORTHERN NEVADA, SEPTEMBER 5, 2020

In September, real life began to invade ski fantasy. This was not unexpected.

During the period that all of this ski training was going down, I was also launching a book, editing a second, writing a third, trying to steer my company through a pandemic, trying to be a good husband, trying to be a good dog dad . . . this list goes on.

Historically, when real life invades ski fantasy, real life wins. This time, I decided to get proactive.

While I'd always been an early riser, a few days ago I decided to push my clock back. Instead of my usual 4:00 a.m. wakeup time, 3:00 a.m. became the new normal, with the occasional 2:00 a.m. rise time if there was a lot going on.

Today, I started testing this schedule. I needed to reset my body's clock before the ski season arrived. I needed to get used to rising early, putting in a five-hour writing session, followed by a ninety-minute weight vest hike, then three more hours of work, a nap, another three hours of work, a trip to the gym, a post-gym recovery activity, then read, watch two SLVSH videos to prime my subconscious, and fall asleep around 8:00 p.m.

Yet this was only the warm-up round. For the next eleven months, I would keep these absurd hours and then some. The weight vest hikes and trips to the gym would disappear, replaced by five days of skiing and an even more unusual schedule.

Twice a week, I would rise at 3:00 a.m. and work until 7:00 p.m.

Twice a week, I would wake at 3:00 a.m. and work until 11:00 a.m., then head to the ski hill, doing work calls along the way. I'd ski until the lifts closed at 4:00 p.m., do more work calls on the return drive, arrive home, and go straight to the mat for a yoga session, followed by a sauna or bath, then ice, then dinner, reading, ski videos, sleep.

Three times a week, I'd wake at 3:00 a.m. and work until 7:30 a.m., then head to the ski hill for a full day session—9:00 to 4:00—followed by work calls on the way back, and the same recovery protocol once I got home.

Why? Flow.

The *challenge-skills balance* is flow's most important trigger. Flow follows focus. The state can only arise when all our attention is locked on the task at hand. And we pay the most attention to a task when its challenge level is *slightly* greater than the skills we bring to it.

But one of my main ideas about peak-performance aging is that the challenge-skills balance shrinks over time. Life leaves scars. As a result, older adults tend to have more subconscious fears than their younger counterparts. These fears, I believe, diminish the size of that challenge-skills sweet spot.

This is why "one inch at a time" has become my motto. I'm operating under the idea that the emotional baggage I bring to athletics, coupled to my injury-riddled past, has shrunk my challenge-skills balance from its normal size to something much smaller. This is also why I introduced my new schedule. It was designed to minimize subconscious fears and maximize the size of that sweet spot.

More specifically, my schedule prioritized fitness and recovery, so energy levels would never be an issue on the mountain. Low energy translates into greater anxiety, and I couldn't take any chances.

More critically, my early-morning writing session always came before any physical exertion.

I don't write well after I exercise, and since writing is how I pay the bills, there was no other choice. The brain treats financial concerns as a basic safety and security need. If I arrived at the ski hill worried about bad writing or, by extension, shaky finances, then those primal fears would further shrivel my already shrunken challenge-skills balance.

Put differently, in the forty-year battle between ski dreams and real life, real life has always won. Always. But not this time.

This time, well, that was the thing: time.

Time was marching on. Time would win this war. The only question was when. Would it win before or after I chased down my ski dreams? *Tick tock.*

This time, I was determined to keep my priorities straight.

## PINE NUT MOUNTAINS, SEPTEMBER 22, 2020

I was six months into my new training regimen. I felt pretty good about my progress. But this wasn't the first time I'd trained hard for a ski season. There have been multiple occasions when my off-season fitness program produced results that looked pretty in the mirror but performed poorly on the hill.

It was time to take some measurements. I headed to the gym and got on the scale. One fifty-six was the number. I'd gained twelve pounds. I got on the bench press. My old max bench was five reps of two hundred pounds; my new max bench was three reps with two hundred and twenty-five pounds.

This caught my attention, as my new max bench was actually my older max bench. Three reps with two hundred and twenty-five pounds was my lifetime best max bench. It was a record set in my late twenties, before I got Lyme disease. The voice in my head had spent thirty years taunting me with this number, telling me that I'd never return to pre-Lyme form, telling me other things. . . .

I took this opportunity to remind the voice in my head of my sixth-grade teacher, a woman who told me that—unless I curbed my reckless behavior—I would never live to see thirty. On my thirtieth birthday I sent her a postcard: "Yo," it read, "I'm thirty. What else you got?"

"Leg press," said the voice in my head.

Leg press would be the real test. The large plates weigh forty-five pounds apiece. In April, my max leg press of three hundred and sixty pounds required eight large plates, four on each side of the machine. Here, in September, I was hoping for five hundred and forty pounds, meaning, adding two additional plates to each side.

In reality, I added those plates, did ten reps, and still had more in the tank. I added two more plates and pushed out six more reps. I added twenty-five pounds to each side, bringing the new total to six hundred and eighty pounds. I pushed out five reps. That was the end of the line. But it was an impressive end. In six months, I'd added twelve pounds to the scale, twenty-five pounds to my bench press, and three hundred and twenty pounds to my leg press.

"Yo, voice in my head," I said. "What else you got?"

"Gym muscles," replied the voice. "Everybody knows—gym muscles ain't real muscles."

## PINE NUT MOUNTAINS, SEPTEMBER 24, 2020

The voice had a point. Every summer over the last six years, I've sought expert advice and adopted a novel workout program, giving it a five-month spin, then measuring the results by a single metric: How many laps can I ski on the first day of the season?

Exhibit A: German Volume Training. I gained ten pounds of muscle. The results looked impressive in the mirror but didn't add up to squat on the hill. Seven laps on day one—my lowest total ever.

Exhibit B: Mark Twight's Insanity. One of the rowdier climbers in history, Mark Twight is the inventor of the fast-and-light approach to mountaineering; founder of the legendary den of workout sadism known as Gym Jones; an old punk rocker, writer, photographer, and fitness guru; the man who provides high alpine, cold weather fitness training and R&D to the Defense Department; the man who trained the three hundred for the movie *300*; the man who trained Batman, Superman, Wonder Woman, and the entire Justice League; and the bastard who designed the most grueling preseason fitness experiment I've yet run. The results: thirteen laps on day one.

Exhibit C: Mike Horn. Among our greatest living explorers, Mike's a solid contender for the "Toughest Man on Earth." Back in the 1990s, armed with a knife and some diving fins, he was first person to swim the Amazon from tip to tail—catching and killing all his food along the way. In 2001, Mike completed a year-and-a-half solo hike-and-sail around the equator. In 2004, he topped that with a two-year solo trek around the Arctic Circle. Then, in 2006, this time with a friend along for company, he walked to the North Pole in the dead of winter—the entire trip taking place in total darkness.

I once asked Mike how he trained for his adventures. He told me he followed a simple protocol. Every few days, he left his home in Switzerland, filled an oil drum with water, rolled the drum to the top of one of the Alps, then carried it back down the mountain on his shoulders. "Do that for a little while," he said, "you're tough enough to do anything."

I did not do that. That shit sounded crazy.

But I did have a way to reliably test my fitness level. I had a thirty-pound weight vest and a boatload of mountains in my backyard. My initial goal had been five days of ninety-minute hikes in a row. I put on the weight vest. Kiko led the way. I came home ninety-seven minutes later. I didn't bother saying anything to the voice in my head.

But still—my silence spoke volumes.

### PINE NUT MOUNTAINS, SEPTEMBER 29, 2020.

Five days, five hikes, and each of them longer than ninety minutes. The weather forecasters said snow would start falling before Thanksgiving. Ski season was six weeks away. I still couldn't slide a rail but, fitness-wise, I felt ready.

### NORTHERN NEVADA, NOVEMBER 15, 2020

Real life was not going quietly. In late September—work stuff, life stuff, other stuff—began to interfere with ski fantasy. By early October, I was stressed out. By mid-October, I was sick. By late October, I was really sick.

I stayed really sick through the start of November. Was this an omen? Was this a ready-made excuse for my future park skiing failures?

Was this the confidence-crushing wedge issue the voice in my head had been waiting for?

Yep. Affirmative. All of that and then some.

Today, in mid-November, I got on the scale and realized I'd lost fifteen pounds. Every ounce of muscle I'd added in the off-season was shed in the final stretch of the preseason. And it was still hot as hell outside. And the American West was still on fire. And the presidential election was still driving everyone mad. What did Artaud say? "We are not free. And the sky can still fall on our heads. And the theater has been created to teach us that first of all."

Suffice to say: Real life won this round.

I did not take it well. I lost my faith in seasons. I lost my faith that snow would ever arrive. I shouted at everyone I knew, even the dogs.

FACTION
FACTION

# Chapter 4

### HEAVENLY VALLEY, NOVEMBER 25, 2020, FIRST DAY ON SNOW

The first day of the season had arrived. On my way up the lift, I told myself this was actually happening: I was on a chairlift. I had skis on my feet. And miracle of miracles—they would soon be sliding downhill.

So what's the big deal with sliding?

Life weighs a lot. In my neighborhood, they've got the gravity set way up high. I've been trying to talk to management about this for years but can't find a way through the red tape. And this is the gift of sliding: Sliding brings lightness; lightness brings forgetting.

The godmother of skiing, Dolores LaChapelle, understood this as well as anyone. "There is an experience of 'nothing' when skiing powder," she writes in her classic book *Deep Powder Snow*. "But the idea of nothingness in our culture is frightening, and we have no words for it. However, in Chinese Taoist thought, it's called 'the fullness of the void' out of which all things come."

What LaChapelle is describing is the experience that scientists call "flow." But I try to save the science for office hours. Out in the wild, I use a different set of terms.

"Fast geometry" is the first of these terms. This is the experience that arises when the body moves at high speeds through a complicated landscape. It's a game of affordances. It's pinball played at thirty miles per hour.

"Gravity dance" is next. When engaged in activities that involve gliding, sliding, bouncing, bounding, brachiating, and such, we are playing with gravity, angles, and speed. At rest, on the surface of the Earth, the human body is a fairly limited machine. At forty-eight inches, basketball god Michael Jordan had one of the highest vertical leaps in history. The world record long jump sits at just over twenty-nine feet. That's the outer edge of the spectrum of possible human motion: a box, twenty-nine feet long and four feet high.

As most of us are not world record holders, our boxes are smaller—say, three feet by ten feet. But this is not true at speed, especially when traveling downhill. Speed, when combined with the slippery angles of a snow-coated mountain, is a doorway into an alternate universe of new possible motions.

Oddly, our experience of moving through this doorway is not singular. We alone are not creating this new possibility space. It's always a cooperative effort. A human being cooperating with primal forces: gravity, weather, mountains. Thus, our experience of sliding, gliding, and bouncing—that is, the by-product of moving at speed through a world of downsloped, snow-coated angles—is always a shared one. A primal cooperation. A gravity dance.

"Ghost Dog" is the last term. When playing fast geometry and engaged in the gravity dance, the "I," as Dolores LaChapelle pointed out, disappears. This is why the weight lifts and the forgetting begins. If I was keeping office hours, I would describe this experience as the front edge of flow, or what's technically known as "transient, localized hypofrontality."

But again, I'm not keeping office hours.

Out here in the wild, when at play with primal forces, the "I" disappears and something else emerges. The primal cooperation, the fullness of the void, or, more simply: Ghost Dog.

And today, on my first day on snow, I was definitely hunting my Ghost Dog.

Unfortunately, it was still early November. There wasn't much snow and only a handful of beginner slopes were open. And crowded.

This also explains why, on my eighth lap of the season, I blasted down Ridge Run, straight-lined past the NO STRAIGHT-LINING sign, and blazed onto a cat track. I was sampling the side hits and airing back onto the flat, when I got cut off by a snowboarder. To avoid an early season near-death experience, I sliced hard right under the TERRAIN PARK sign.

The sign was oversized, the shadow it cast enormous. Suddenly, the light went flat. The shadow hid a roller.

"Oh shit," said the voice in my head.

I blindsided the roller, bounced into the air, supermanned for twenty feet, smashed chest-first into the ground, and slid for another fifteen, finally coming to a rest at the feet of a young couple. To cover my embarrassment, I looked up at the couple and shouted "Safe!" as if I had just stolen a base.

They laughed. I laughed. But I wasn't safe. My right rotator cuff was torn. I was going to have to deal with the pain for the rest of the season.

I did sixteen laps anyway. I found my Ghost Dog around lap twelve, and my foul mood lifted. It happened so quickly, it felt like a mirage. Then I beat my previous first day's best by an extra three laps.

So, maybe, not a mirage

### HEAVENLY VALLEY, NOVEMBER 26, 2020, SECOND DAY ON SNOW

Today marked the first of the right butt cheek crashes. I was riding my brand-new Faction CT 2.0s, and, before we get to the butt cheek crashes, let's start with the nomenclature.

The Faction in "Faction CT 2.0" is the name of the company, technically the Faction Collective. The CT stands for ski legend Candide Thovex, which is to say, the CT is a tool designed by CT to do what CT does best: go very fast everywhere on the mountain, throw outrageous tricks anywhere on the mountain.

The 2.0 refers to the model, which refers to the width of the ski beneath the sole of the foot, or what skiers mean when they ask the question: "Sick kicks, brah, what's it underfoot?"

My Faction CT 2.0s are 102 millimeters underfoot, putting them in the low end of the midfat range, or what's sometimes described as a "daily driver"—a ski that you can bring out every day because it's ready to meet whatever snow conditions the day delivers.

Model-wise, there's also a skinnier version of the CT that's 92 millimeters underfoot and creatively named the 1.0. This model is designed for East Coast ice skiers, or terrain park skiers, or both. There's also the 3.0, a fatter, 120-millimeter version that's good for powder days, or the crud-busting chop-bashing required for post-powder days. Finally, there's a 130-millimeter superfat ski, which is designed for Japan, British Columbia, and other meccas of the superdeep, bring-a-snorkel pow-pow.

If you're wondering why you would bring a snorkel to mecca, it's so you can breathe in the white room. If you're wondering what the white room is, well, we're talking about skiing through snow so deep that it quite literally takes your breath away.

The CTs and I did two warm-up laps on Maggie's. Ambition set in fast and I decided to try my first sliding spin 360 of the season. There were high banks on the left side of the run that looked perfect for the trick. Still, I'd only spun a 360 on a banked wall twice before. Once at Kirkwood at the end of last season, once at Mount Hood over the summer. Both times the trick emerged organically: I saw the terrain feature, my body knew what to do, and I did the trick, long before my conscious mind had time to get involved.

Today, I saw those high banks from a hundred yards out and started thinking. I was thinking too much. The voice in my head was stuck on, "Nah, nope, not today."

But I did it anyway.

The problem was speed. Candide Thovex skis the mountain at full Mach—think forty-to-sixty miles per hour. On my first spin, at somewhere near fifteen miles per hour, I discovered the CTs weren't built for slow-speed rotations. The edges felt sticky. The tails felt long.

On my third attempt at a sliding spin, I discovered it was more than a feeling. My tails caught as I was twirling from backwards to forward. Ground, meet Steven, Steven, meet ground. My right butt cheek took the brunt of the impact.

A little later, I caught my tails again. One moment, I was bashing through the bumps, the next moment I was slamming into hard earth. I lost my skis—so, good, the new bindings actually work—and smashed my right butt cheek again. Bone bruise. I would be black and blue for the next two months.

I did another sixteen laps anyway.

### KIRKWOOD, NOVEMBER 28, 2020, THIRD DAY ON SNOW

I took a day off to rest, then got after it again. Skiing three out of four days would be a solid start to my season. The last time I was strong enough to do that was before I got Lyme disease—some twenty-three years ago.

That said, today was my first day at Kirkwood, and this would be the more strenuous test. Even the drive to Kirkwood is strenuous.

Getting to Kirkwood requires navigating a pair of treacherous mountain passes that are prone to freakish weather. Blizzards park themselves overhead for days. Avalanches trap drivers in their vehicles overnight. This winter, whenever I drove to Kirkwood, I brought along an emergency medical kit, extra clothing, heavy snow boots, a flashlight, shovel, two wool blankets, one space blanket, and three days' worth of food and water.

Also, calling Kirkwood a ski resort is an overstatement. Besides the mountain, there's not much going on, and none of it fancy. A general store, a ski shop, a cafeteria, a pizza joint, one sit-down restaurant, a couple of bars, and little else.

The mountain is the real draw. Its terrain is ferocious. Chair 10—aka "the Wall"—routinely makes lists of the "Hardest Runs in America." Just getting on the lift requires pushing past a large skull-and-crossbones flag. This isn't false bravado. The flag is actually the ski patrol trying to talk tourists out of false bravado. Chair 10 may be the only chairlift in America where lift operators routinely try to dissuade customers from taking the ride. Or, as I said to Ryan all last year: "This cannot be legal."

And today, it wasn't. The resort was still two snowstorms away from being able to open the upper portion of the mountain. Instead, we stayed lower mountain, lapping low-angle slopes and looking for ways to make our laps interesting.

It didn't take long. On our second run, Ryan noticed a three-foot stump wedged between two tall trees. No way was Ryan going to jump off that thing on our second run.

An instant later, Ryan jumped the stump, nailed the landing, and skied away. "If Ryan can do it, so can you," said the voice in my head.

"Fuck off," I replied.

And that's when it dawned on me: In order to have the season I wanted, I would have to do things that scared me on a regular basis. This meant that the fear that I was now feeling I was going to be feeling for a long time to come.

But the rules were the rules.

So, while trying to match my speed to Ryan's, I jumped the stump, stomped the landing, and skied away with a big smile on my face. I knew what this meant. Like my wall ride 540 at Mount Hood, clearing that stump jump was more proof of concept.

Then I jumped every other damn jump, stump or otherwise, that Ryan found today—because, well, you can't trust a concept.

## HEAVENLY VALLEY, DECEMBER 3, 2020, FOURTH DAY ON SNOW

Fourth day of the season and I was already sore. But I walked Kiko, did yoga, and felt better. Still, I went to the Heave—which is what the locals call Heavenly Valley—with low expectations.

I exceeded those expectations: skied eighteen laps, including three tentative laps backwards, in my first attempt to learn to ski switch.

So, what's with the lap count?

It comes down to goals. Humans perform best with three levels of goals: mission-level goals, high-hard goals, and clear goals. Mission-level goals are lifetime goals. High-hard goals are the multiyear achievements required for those lifetime goals. Clear goals are the daily actions needed to accomplish those high-hard goals. If all three levels are pointed in the same direction, and progress toward those goals remains steady, then motivation and momentum—aka dopamine and flow—are the result.

In this case, my mission was to advance flow science. My season was a high-flow experiment in peak-performance aging. This meant every time I went to the mountain, I'd get a little dopamine for showing up and advancing the cause.

Could I learn to park ski at age fifty-three? That was my high-hard goal. Progress toward this goal would be measured by my ability to learn the tricks on my trick list. So, every time I tried a trick, I got a little dopamine for taking a risk and trying something new. If I actually pulled one off successfully, I got a lot of dopamine. Since a lot of dopamine significantly increases focus, I dropped into flow as the result.

Yet there were outside factors in play—injury, weather conditions, COVID flareups—that I couldn't control. As a result, I needed clear goals, or the goals that would fill out my daily to-do list. These clear goals would benchmark my progress, yet they needed to have enough flexibility to handle changing conditions on the ground.

For example, if I arrived at the mountain to discover ten inches of powder and soft landings, then sure, trying a 360 is fine. But if there was sheet ice everywhere, trying a 360 shouldn't even be a consideration. I didn't want to pressure myself into stupid decisions, not if I wanted to stay out of the hospital. Thus, for my clear goals, I needed a way to measure the quality of a ski day that was unrelated to my terrain park and big mountain progress.

This was why my lap count mattered.

Twelve laps was a mellow workout. If conditions were terrible, or I was exhausted, twelve laps was my minimal acceptable goal. It meant I was maintaining fitness, but not overtaxing the system. Done correctly, a little dopamine might come my way.

Sixteen laps was a solid workout. Barring bad weather or nagging injury, sixteen laps was my standard "good day on the hill" requirement. It meant I was doing the work and advancing the cause, even if, on the day in question, conditions weren't right for me to ski a line on my line list or learn a trick on my trick list.

Twenty-plus laps was a great day. It meant I was skiing strong, advancing my fitness, and should conditions allow, I'd be ready to attempt one of the harder lines or bigger tricks on my lists. Twenty-plus laps was also a guarantee that I'd get into flow along the way, as it's too tiring to ski that much without the pain relief that comes with the state.

Put differently, when I woke up today, I felt pretty beat-up. Thus, my goal was a twelve-lap day. By skiing sixteen laps, I exceeded my expectations, got a little dopamine for reaching my initial goal, got a little more for reaching my stretch goal, and left the mountain with more motivation than I had when I arrived. This creates momentum.

On the way home, I spent that momentum on errands. I was in need of the necessities. I stopped off at the pet store for dog food, the grocery store for human food, and the weed store for Steven food. Then I went home and took a bath, adding half a bag of Epsom salt to the tub.

As I said, I was already sore.

## KIRKWOOD, DECEMBER 5, 2020, FIFTH DAY ON SNOW

As LL Cool J said, "Game recognize game."

In this game, on the second run of our day, Ryan and I caught a flash of a guy in a blue ski instructor jacket catching big air off some invisible feature on the side of the hill, then sailing into thick woods.

Game recognized game. Game on.

Ryan and I took off after the ski instructor. We didn't really know our way around Kirkwood. We just knew there were awesome secrets everywhere, if only someone would show us where to look.

It was awesome, all right. Later, we'd call this line "Stump Run"; though, at this point we didn't know what was coming, but it was coming at high speeds.

Ryan spotted the invisible feature first. It was a ramp of snow rising over a fallen log at the edge of the woods. The takeoff looked perfect—but where the hell was the landing?

I glanced right, I glanced left, all I saw were big clusters of pine trees.

Ryan must have seen something I missed. He was about twenty feet ahead of me and then he wasn't. He hit the ramp, aired into the woods, then vanished. I didn't hear a crash so, apparently, game still on.

I hit the ramp and sailed into the air. "A-ha," said the voice in my head, "the landing isn't the problem."

The problem was the very large pine tree three feet in front of the landing. The predicament required what used to be called a "McConkey turn," because there was no one better than the late, great Shane McConkey at executing high-speed slashes of the save-your-ass variety.

Ass saved.

That first kicker led to a line of them: a log jump into another log jump into an Evel Knievel ramp over a huge stump into a just plain evil ramp over another stump. I cleared the first log, cleared the second log, watched Ryan air to the moon off the huge stump, and decided to pass. Then I hated myself for passing and decided to redeem myself on the last of the stumps.

Redemption shot me fifteen feet up and thirty feet forward. Normally, this was when I would panic. Not this time. This time, I knew what to do. I sucked my knees toward my chest, recentered in the air, and laced the landing.

As I skied away, I caught sight of the same blue-jacketed ski instructor. He was now parked by the side of the trail. He gave me a tiny pole clap as I passed—maybe the first time in my life anyone has ever clapped for anything I've done on skis.

Stump Run kicked us into group flow, and we flowed right into a game of "Chase the Rabbit."

Chase the Rabbit is a game of progression, played with mirror neurons and high speeds. The way it works is simple: someone leads, someone chases. If Ryan's the rabbit, for example, he takes off down a trail and I tuck in behind. My goal is to do everything he does or, occasionally, something better.

Chase the Rabbit amplifies progression like little else, especially if you're playing in flow. It comes down to biology. Humans are hardwired for imitation. We're "great apes," and there's a reason—aping, copying, mimicking is how we learn. Since learning is fundamental to survival, our aping happens automatically.

In flow, both self-awareness and fast-twitch muscle response are heightened. We're better at tuning in to the body's subtler signals, and better at acting on those signals. And one signal matters more than most.

During Chase the Rabbit, milliseconds after the leader makes a move, the follower's mirror neurons run the same motor program—simulating the leader's move in the follower's mind before the follower has to execute that move in reality. In the next millisecond, the brain's pattern-recognition machinery, via the experience we call "intuitive self-awareness," tells the follower—either with a tiny *pleasurable* squirt of dopamine or a tiny *fearful* squirt of norepinephrine—if they have the skills to execute that move. It's not a two-step process: monkey see, monkey do. Rather, there's a third step: monkey see, monkey get a signal if monkey can do, then monkey do.

As long as you get the green light, listen to that signal, and never hesitate in your execution, you have the skills to make the move. Hesitation is death. It amplifies fear. In response, muscles grow rigid in preparation for a possible crash, and conscious awareness roars back online to help prevent that crash.

Unfortunately, conscious awareness blocks the automatic execution of the motor patterns needed to prevent that crash. In other words, all day long and all season long, monkey see, monkey check the signal, monkey do, and onward toward the promised land.

## KIRKWOOD, DECEMBER 9, 2020, SIXTH DAY ON SNOW

The Draw is a long gulley filled with rollers, rocks, and cliffs. The walls are steep and tall and dotted with a series of increasingly large side hits. Last year, I tested the six-foot starter ramp and the twelve-foot intermediate ramp, but stayed away from the thirty-foot motorcycle booter. The last seven feet of that booter were completely vertical—which made it perfect for a backflip.

No, thank you.

If you're not interested in a backflip, a "shifty" is the safer option. During a shifty, the skier twists their lower body in one direction and their upper body in the opposite direction. Beyond style points, a shifty gives the body something to do midflight and prevents "dead sailoring"—that is, a midflight rigidity that can tilt a skier off-axis and produce, well, dead sailors.

I didn't know how to shifty, but I knew how to throw a "mule kick"—which is sort of an off-axis shifty. Like my sliding spin 360, the mule kick was an already ingrained movement pattern that I could execute with little fear and a high chance of success. And if I toned down my mule kick into a mule twist, I figured, with a couple dozen repetitions, I might be too sexy for my shirt.

I tried to shifty off the motorcycle booter. The results: lukewarm. I would keep my shirt.

But the day wasn't over.

We left the motorcycle booter and skied toward the other side of the resort. I spotted a small mound to the left of the cat track.

"Nose butter 360," said the voice in my head.

I'd never attempted a nose butter 360 before, but I'd been given the "go" signal and there was no time to argue.

Once again, the results were lukewarm, but, you know: *Fuck yeah, lukewarm!*

Thirty seconds later, Ryan threw a 180 off the small lip formed by the roots of a cedar tree. My good mood evaporated. Fear took its place. This was the gap in my game, the small hitch in my big plans for park skiing domination. I had no idea how to throw a 180.

After drawing up my goal list, I knew I'd eventually have to learn to throw a 180. Yet, until now, I'd suppressed that knowledge, since learning a 180 required learning how to land backwards. I couldn't imagine any way that landing backwards wouldn't immediately become backwards, sideways, and otherwise.

But Ryan threw that 180, and the rules are the rules.

*Holy crap,* is landing backwards an amazing feeling. My brain didn't believe it was possible, but my body understood the motion perfectly. I landed backwards, absorbed the shock with my knees, and, surprisingly, knew exactly what to do next: I spun forward like I was executing the second half of a sliding spin 360.

I'd just learned two tricks in under a minute. I'd never learned two tricks in a year before. I felt like I'd been shot through a wormhole and was currently violating the space-time continuum.

Up until now, my one-inch-at-a-time plan of park skiing attack was either an unproven hypothesis or a midlife crisis, depending on who you asked.

Not anymore.

"Happiness," said the voice in my head. "That strange tingly sensation you're feeling—humans call that 'happiness.'"

## HEAVENLY VALLEY, DECEMBER 16, 2020, SEVENTH DAY ON SNOW

We skied with Ryan's younger brother, a serious endurance athlete with a damn good park game. What I remember most was moving at silly speeds through old-growth forests, airing off logs and stumps and who knows what else. It was all a blur of fast geometry. Yet what I liked best about our approach was the privacy.

I prefer to do my skiing out of sight, and this was especially true when attempting to learn anything new. I'm hard on myself under normal conditions. If I add public embarrassment to those conditions, I can get stuck in a shame loop for days. This is another secret to making progress in Gnar Country—know who you are and how you like to learn, and don't waver.

Today, I didn't waver.

We skied lap after lap, hidden in the trees, attempting to turn the entire mountain into a slopestyle course. For those unfamiliar, "slopestyle" refers to a freeskiing discipline in which athletes compete on courses that mix jumps, rails, wall rides, and such. When turning the mountain into a slopestyle course, our goal was to jump over, slide across, and twirl down as many features as possible.

Today, for example, my favorite section involved a quick ollie over a manzanita bush to jump out of the woods, a nose butter 360 over the knoll in the center of the trail, then a switch snow slide across a small berm to set up a soaring log jump back into the forest.

At the heart of our slopestyle game was the search for a safer path to progression, but this is best explained by example.

Earlier, I mentioned my trip to Mount Hood with ski legend Glen Plake. On that same trip, Glen and I hiked above Mount Hood's glacier to ski one of the nastier chutes up top—steep, cliff-lined, definitely not a place to play around.

Yet, before dropping in, Glen backed up about thirty feet, skated hard toward the chute, aired off a mogul, and executed an "airplane turn"—that is, a midair ninety-degree rotation—to set up his line.

Like, what the hell?

After I skied the chute, I had questions. "Hey, Glen," I asked. "Why do something dumb and dangerous before you do something dumb and dangerous? I mean, doesn't that just ramp up the odds of disaster?"

It's been about twenty years, but Glen said something like: "You don't get it. When I threw that airplane turn, it dropped me into the zone. Being in the zone gives me the best chance of surviving the dumb and dangerous."

And Glen wasn't wrong.

When he first saw the mogul and realized he could use it to throw an airplane turn, that was pattern recognition. So important is this ability for survival, the brain rewards pattern recognition with a squirt of dopamine. Additionally, an airplane turn also creates a moment of weightlessness: a novel and unpredictable sensation that pushes even more dopamine into the system.

In Glen's case, all this dopamine added up into flow—meaning, before he started skiing the dumb and dangerous, he used that airplane turn to put himself in the exact peak-performance state required to survive the dumb and dangerous.

Good thinking, Glen.

Younger athletes don't think this way. They often use risk and the dopamine produced by risk as their primary flow trigger. This is the "just send it, brah" attitude that gives action sports their adolescent flavor. It's also an attitude that works fine for talented athletes, but it's not fine for me—that is, older, not talented, not interested in any more serious injuries.

Instead, my goal was Glen's goal: I wanted to use creativity—that is, pattern recognition, novelty, and unpredictability—as my gateway into flow. This is why we turned the entire mountain into a slopestyle course. It forces the brain to make creative decision after creative decision after creative decision. This piles up the neurochemistry, providing a safer and reliable entrance into the zone.

This season, the goal was to use our slopestyle game to drop into flow and, once there, head into the park to learn new tricks, or head into the steeps to push our skills. Trying to do this in reverse—that was the very thing that sent me to the hospital on so many prior occasions.

Today, no one went to the hospital. Instead, oddly, Ryan and I dusted his brother after twelve laps. The well-trained endurance athlete in his late twenties dropped out and went home. The fifty-three-year-old man did ten more laps with Ryan.

All of those laps were spent in flow. One of them took us off a little rock drop beside Comet Express. It wasn't much: a five-foot cliff requiring ten feet of air, but you had to make a midair airplane turn or risk hitting a tree. The year prior, I'd shied away from this drop.

This time, we skied over to take a look. Ryan liked the look. He hopped off the cliff and the rules are the rules.

Yet because I was already in flow from our slopestyle adventures, my first cliff huck of the season led to my favorite feeling in the world: much ado about nothing. The huck was simple, the landing smooth, and the confidence that resulted—now, that's the real jet fuel of progression.

### KIRKWOOD, DECEMBER 19, 2020, EIGHTH DAY ON SNOW

We got to Kirkwood early. Enough new snow had fallen so that Chair 6 was running, giving us access to the upper mountain for the first time this season. It wasn't the whole of the upper mountain, but I would get to ski Oops and Poops again, site of last season's sliding spin 360.

All off-season, I'd been thinking about my return trip to Oops. Expectations were high. I assumed there would be a jolt of pattern recognition—that I'd see the spot where I'd thrown that sliding spin and recognize it again. Of course, my real fantasy was that I'd see the spot, recognize it again, and sliding spin 360, no problem.

But it was a problem.

Just the possibility that I might have to reexecute the move raised my stress level. In the language of flow science, this knocked me out of the challenge-skills sweet spot and blocked my entrance into the zone. In the language of teenage boys, smooth move, Ex-Lax.

Meanwhile, in the language of motor learning, we acquire physical skills in three stages: *cognitive*, when we have to think through what we're doing; *associative*, where we are starting to learn all the patterns of the movement and starting to turn them into the more complex motor patterns known as chunks or schemas; and *automatic*, where we can automatically execute these complex motor patterns with no cognitive interference. The issue is that fear often kicks you down a stage.

My ability to execute a sliding spin 360 in a natural halfpipe was in the early portion of the associative stage. My anticipation, my expectations, my Gnar Country for Old Men dreams, all of it increased my fear, boosting norepinephrine, blocking flow, and worse: kicking me down a stage. So, yeah, I was back to cognitive.

Finally, in the language of neuroscience, there's a different issue at play: repetition suppression. Novel stimuli catch our attention. But when stimuli are repeated, each repetition produces a smaller neurological reaction. Repetition of stimuli suppresses our reaction to that stimuli—aka familiarity breeds indifference.

Put differently, if you've never visited the Grand Canyon, when you arrive, you can't stop staring at that amazing hole in the ground. Four days into that trip, the hole becomes that damn thing you have to walk around on your way to the gift shop. In other words, upon repeated exposure, the brain reduces the wow factor, because, neurobiologically, the wow is expensive to produce. The wow burns calories, which defies the brain's basic need for efficiency.

One difference between good and bad athletes is how long it takes to arrive at repetition suppression. Ryan is a good athlete. If he tries a trick once and executes it successfully, he can repeat the feat with little fear and no problem. Once is enough for him to create the associative memory—that is, the schema or complex motor pattern—required to do it again. As a result, the next time he tries, there's repetition suppression. This translates into less fear, more flow.

Unfortunately, that's not the case with me.

Over the past decade, I've had three major surgeries and six so-called procedures, and that doesn't include the two broken ribs and one collarbone fracture that were never treated. Ryan can learn a trick in a single go. My injury history jacked up my fear levels. Often, I have to execute a move five times or more to get to the edge of associative learning and the beginning of repetition suppression.

But whatever, we all have our shit.

In my case, I got my shit all over Oops and Poops today. And with a name like Oops and Poops, I definitely should have seen that shit coming.

## KIRKWOOD, DECEMBER 23, 2020, NINTH DAY ON SNOW

They opened the rest of the upper mountain and we got to ski the Wall for the first time all season. We took our "hello again" lap down Schafer's Chute, a "we missed you" lap down Sisters' Chute, then made our way to All the Way Chute, which is maybe my favorite chute in the world.

When COVID closed the resorts, it was All the Way I missed the most. I pined for it like a teenager with a crush. I couldn't stop thinking about this line, how this line made me feel, and all the naughty things this line and I could do when we saw each other again. "So, hey, All the Way, can I get this dance?"

Finally, the Notch: a steep and narrow natural halfpipe that is sort of like the Oops and Poops halfpipe, except meaner.

Even the entrance can be spicy. A monster wind lip divides All the Way from the Notch. It's a one-hundred-foot-long, twenty-foot-high wall of snow. Today, Ryan jumped off wall, carved onto the far side of the chute, then carried that speed straight back toward the wind lip—and what the hell was he thinking?

He was thinking sliding spin 360.

My jaw dropped. Five days earlier, Ryan had started practicing sliding spins for the very first time. This was on flat ground. Two days ago, he'd added some forward lean to the motion and attempted his first nose butter 360. Today, Ryan spun a sliding spin 360 across a completely vertical section of wind lip wall—which is to say, repetition suppression on the job.

Now it was my turn.

I didn't want anything to do with that wind lip, so I skied farther down the run and threw a sliding spin 360 on the chute's less vertical left wall. It wasn't nearly as steazy as Ryan's effort, but the last time I'd tried the trick was last year, in the Oops's halfpipe, and then I'd only managed to spin a 270.

Today, I beat that record by ninety degrees and called it a win.

We rode that win back for another lap. This time, Ryan tried a nose butter 360 off the top of the wind lip. Ambitious. Too ambitious. What neither of us had noticed? The temperatures had warmed and the snow was sticky.

Ryan spun two-thirds of the way around, lost speed, caught an edge, and tumbled down the wall. He cartwheeled once, backflipped twice, and almost snapped his ankle. He skied away from the crash, but his ankle would ache for the rest of the season.

Speaking of things that would ache for the rest of the season, I then had the year's first major encounter with vertigo.

On that same second trip down the Notch, about two minutes after Ryan's fall and two-thirds of the way down the chute, I caught a glimpse of the blue-jacketed ski instructor—the same guy who had led us to Stump Run.

Chase the Rabbit. Game on.

I followed him into the woods. We zagged left. We zigged right. We shot back onto the run and back into the trees and suddenly the blue jacket was gone—and where the hell was I?

I sliced out of the woods and realized I was perched atop a knife-edged ridge, with heavy exposure in all directions. I couldn't see the ski instructor. I couldn't see the bottom of the run. That's when it clicked—I was surrounded by cliffs.

What does the onset of severe vertigo feel like? Like a hostile takeover by an angry four-year-old, with the decision-making skills to match.

And I had a decision to make: The only way out was either a blind cliff huck or a kick turn on a knife-edged ridge. In my condition, the cliff was too big a risk. But a kick turn on a knife-edged ridge requires balance. My balance had already split town and the spinning was about to start. I had a second to move.

I got angry. I was mad at myself for being afraid. Pissed at the universe for creating the emotion of fear. And seriously ticked that the cowardice in my past was still turning into shame in my present. The result: I roared. In fifty years of skiing, not once have I roared in response to vertigo—but, Holy Biology, Batman!—that roar worked.

The roar produced testosterone, which counteracted the vertigo. Muscle control came back immediately. I jumped into the air, twisted a 180, landed back on the ridge, and skied the hell out of there—except, there was nowhere to go.

The only exit was a steep, straight line between two large rock walls. Straight lines were worse than blind cliffs. I had a tiny bit of experience with blind cliffs. I had zero experience with steep, straight lines. Why did I freak out so badly after making my "line list" for the season? Because five of the lines on that list had straight-line exits.

But I was already in fight mode and roared again, this time intentionally, just to see if it would work again.

It worked again. I nailed my first straight line of the season. It was another win. Then I got off the hill before the voice in my head had time to argue.

### HEAVENLY VALLEY, DECEMBER 24, 2020, TENTH DAY ON SNOW

When I see people skiing in furry animal costumes it always reminds me of the generation gap. In my day, the only time we dressed up in furry animal costumes was for all-night, drug-fueled orgies. Today's generation—they think it's just fine to wear them while skiing and snowboarding.

### KIRKWOOD, DECEMBER 26, 2020, ELEVENTH DAY ON SNOW

It was the first big blizzard of the season and the Kirkwood 500 was in full effect. Black ice on the highway, whiteout conditions, and every powder junkie in the state barreling toward the resort at maximum velocity.

The hell drive freaked me out. So while all the other powder junkies rushed to the upper mountain for big drops and steep chutes, I stayed on the lower mountain, skiing mellow tree runs until I calmed the freak down.

Then I got my freak on.

Five laps, ten laps, twenty laps, more. It was all fun and games up until the moment it wasn't—which is the way it goes at Kirkwood.

Ryan and I decided to end our day on Chamoix, which is the chute that sits directly beside Oops and Poops. Later in the season, the snow would fill in the upper portion of Chamoix, and some of Kirkwood's biggest lines would be in play. Today, with bare rock above us, we traversed below the top and cut in from the side, aka we made a hard left turn above a death-fall cliff, then prayed for the best as we took a hard right turn and dropped over a rock and into the belly of the chute.

But not so fast. Without much snow, what was usually a four-foot rock drop had become an eight-foot fall. I went from high speed to full Mach. Instead of flowy turns in familiar places, I was doing the rock dance through uncharted territory.

Three turns later, it got worse, as three turns later, I ran out of snow.

Chamoix is a funnel. The choke point is a six-foot gap between two rock walls. Normally, the exit requires a NASCAR turn through these walls. Today, the exit was a sheer cliff of bare rock. The only way out was to jump the cliff, or what skiers call a "mandatory air," and what I have come to know as "instant vertigo."

I hockey stopped above the rock. I started to twitch. I twitched over to the far right side of the chute for a better look. I didn't like the look and the world began to spin. I tried to kick turn and fell on my ass. I got back on my feet, took a deep breath, and slid to the far left side for a different look. This side required a sliding right turn on a tiny blade of snow, and then a mandatory air over a six-foot section of rock.

But something unusual happened: I relaxed. Then I realized why.

That sliding turn—it looked familiar. My body understood the motion. Hard carves on thin blades of snow were what passed for spring skiing in New Mexico. I had years of practice making that turn. And a six-foot cliff would produce about twelve feet of air, and I jumped farther than that all the time.

Suddenly, I knew what to do. My brain took two automatized movement patterns—a hard right turn on a thin blade of snow and a twelve-foot air—and linked them together into a new pattern, a larger pattern, or what psychologists call a "chunk" and what motor-learning experts term a "schema." It was a moment of pattern recognition that pushed dopamine into my system.

This explains another rule: See the line, ski the line.

The dopamine window produced by pattern recognition doesn't stay open for long. Inside the window, courage and fast-twitch muscle responses are amped way up, but you have to go-go-go before it's gone-gone-gone.

The drop was bigger than I expected, but I stuck the landing and skied away and, once again, roared. Ryan thinks this cliff and that roar were the moment everything changed, the moment my skiing changed forever. I didn't know that yet. What I did know—I wasn't in the hospital and I wasn't filled with shame, and every other time I've encountered a mandatory air, those have been the results. Plus, we'd done nearly thirty powder laps, which meant I was in midseason form and it was still early season.

No question about it, I was fifty-three years old and in the very best shape of my life. If I wanted to turn my Gnar Country experiment into the next Abs of Steel fitness program—well, see for yourself, that shit would sell like hotcakes.

# Chapter 5

## NORTHSTAR, DECEMBER 27, 2020, TWELFTH DAY ON SNOW

I went to Northstar for the first time all season—and what was I thinking?

With the Christmas holiday in full swing, the crowds were everywhere. I hid in the trees on the mountain's backside. Incredible views, but the terrain is too mellow for much more than cruise laps and dumb airs. I spent the morning going fat to flat, and my shins felt it later.

By afternoon, I decided to get to work. I hadn't come to Northstar to ski the trees on the backside. I'd come for the terrain park—the primary reason people come to Northstar.

Challenge-wise, Northstar is a beginner mountain. The resort makes up for this deficit by building one of the bigger terrain parks in Tahoe. By midseason, a razzle-dazzle of features would fill the trails. There would be beginner lines, progression parks, and expert fare. Unfortunately, all of this required snow. To date, not enough had fallen for anything beyond a four-pack of medium-size jumps and a handful of rails.

You could say I attempted my first jump line of the season on that four-pack of jumps, but that would be overstating the case. After failing to clear the initial two tabletops, I decided to avoid the third entirely. The final jump was a large kicker built atop a larger knuckle—that is, a small hill with a flat crest and a steep backside. I'd been watching SLVSH skiers throw tricks off knuckles for months. I had big plans for this feature. So I skied up the knuckle and peered down the backside and felt all the air rush out of my season. . . .

It was huge and steep and way out of my league. Four jumps, four fails—an inauspicious start to my park skiing career.

Then the jump line became the rail line, and things went from bad to worse. I saw forty-foot double-kink flat bars and twenty-foot lipstick tubes and nothing but a long dancer box for beginners like me to slide.

The dancer box was a long rectangle of white plastic laid across the surface of the snow. I decided to attempt my first "surface swap 360"—which is a sliding spin 360 except, instead of being done on the snow, it's executed on a surface: a box, rail, or wall ride.

I was pretty sure I had the trick on lock. It used the exact same motion as a sliding spin 360. All I had to do was ski onto the box and twirl.

I skied onto the box, started to twirl, and twirled onto my ass. My legs shot left, my torso shot right, my already injured right butt cheek took the brunt of the impact. I slid the rest of the dancer box on my side—with about three hundred people watching.

Welcome to park skiing.

## HEAVENLY VALLEY, DECEMBER 28, 2020, THIRTEENTH DAY ON SNOW

I woke up sore, but yesterday's public ass kicking made me crave a little private redemption, which meant a solo mission was in order.

I was still sore when I got to Heavenly, but decided to suspend judgment about the day's potential until after I'd completed my standard four-lap warm-up. Lap one was a fast cruiser to get my body used to being in motion. Lap two was a faster cruiser to get my brain used to the speed of skiing. Lap three began as a tree run and ended as a bump run and both were aerobically taxing—which was the point of lap three.

Lap four was my response to the unpleasantness of lap three. I like to ski to music, but never until this fourth lap. Before then, my body isn't awake enough to move to the beat, and music hurts more than it helps. But once warmed up—aka run four—I can catch the rhythm.

I also caught air—which was the point of lap four. I didn't jump off anything big. Just big enough to give me that the feeling of flight and the dopamine it produced.

Once I had that feeling, I was ready to roll. In this case, I rolled right into an old game: Pass the Fast. On the chairlift, I tried to find the fastest person on the hill. Once I got off the lift, I'd try to catch and pass that person on the way down.

Today's target was male, roughly age thirty, in a baggy beige jacket, a pair of Armada skis, and a three-hundred-yard head start. It took me almost the entire run to catch him, finally sailing past on the last section of bumps that led back to the lift.

But he had caught on to the game.

On the next run, I had the head start and he passed me midway down the slope. Just so I knew that he knew, he shiftied off a little ledge directly in front of me, giving me a little ski-wag before his beige jacket vanished into the woods.

Not so fast, Baggy Beige.

I caught him in the woods, which, admittedly, is a creepy thing to do under other circumstances, but in this circumstance, good skiers like chasing good skiers because that's how they become better skiers. Thus, I passed him in the woods.

By the time we got back to the lift, we were laughing about the game. We took four laps together. By the end of those laps, I was in

Ghost Dog mode, which was the head clearing I'd come for. But I didn't want to push it—so goodbye to my new friend, back to the car, and home for a nap.

### KIRKWOOD, DECEMBER 30, 2020, FOURTEENTH DAY ON SNOW

Every time I drive past a church in the mountains, it makes me wonder about people who build churches in the mountains. Each of them stood in God's cathedral and thought, what? That a parking lot and some hammered boards could improve matters?

### KIRKWOOD, JANUARY 3, 2021, FIFTEENTH DAY ON SNOW

So far this season, I had yet to attempt any of the lines on my line list. It was time to change that fact.

I decided to start with "the Fingers," a series of five chutes carved into a larger cliff that occupies the center of Kirkwood's aptly named Cliff Chute. Each finger is separated from the next by a blade of lava rock. From afar, the rocks look like knuckles, the chutes look like fingers, and their names—Eeny, Meeny, Miny, Moe, and Main Finger—reflect this fact.

Ryan and I had stared at the Fingers all last season. They were some of the most aesthetically enticing chutes I'd ever seen. A foreboding off-world landscape lifted from a trippy 1970s album cover. Could we ski through the trippy? That was an open question.

There were a few Go-Pro videos online, but the camera flattened the terrain. What looked mellow was actually steep; what looked steep was actually insane. The Fingers looked steep. And narrow. With lava rock walls, a bad fall could mean a trip to the emergency room. Plus, all the lines had mandatory airs or straight-line exits. Could I ski those lines and make those exits?

Last summer, before I had answers to those questions, I put Eeny,

Miny, and Moe on my line list. It was a leap of faith. So far, lack of snow had kept me from having to test my faith. But today was day fifteen, and lack of snow was no longer my problem.

Lack of courage was now my problem. I was going to have to boldly go where Steven had never gone before. No way was I going alone. I dragged Ryan along for the ride.

Actually, I made Ryan take point. It helps fight the vertigo. If I stay on Ryan's tail and keep my eyes locked on his skis, never pausing, never giving the exposure a chance to bitch slap me into the stratosphere, I had a chance.

I wanted that chance. I was hell-bent on Miny. In videos, the chute ran beneath an enormous rock overhang—like skiing through a cave. In person, as we stood at the top, the cave looked steeper, narrower, and with a way more ferocious straight-line exit than anything I'd seen on video.

But a funny thing happened as I watched Ryan ski: I recognized the moves. The fast right turn into the chute made sense. The second and third turn as well. The fourth and fifth each required hard slashes in high consequence zones, but they too were within my skill set. Yet the straight-line exit—that made no sense whatsoever.

Still, I'd straight-lined out of the Notch; how much worse could this be?

The first three turns were as expected, the fourth and fifth were both flowier than expected. Then I discovered a choice: Either aim directly for the exit—a forty-foot straight line and too spicy for me—or carve high on the left rock wall and try to exit from there.

The left wall looked less intimidating. It was close to fifty degrees steep yet made the exit appear a straighter shot. It also, as I discovered, made the exit completely blind. One thing remains true: If you can't see where you're going, it's always a little surprising when you get there alive.

Then we skied Miny three more times to make sure it stuck.

## KIRKWOOD, JANUARY 6, 2021, SIXTEENTH DAY ON SNOW

I came back to Kirkwood alone. Skiing a heavy line with Ryan is one thing; a solo mission is another. But I had to know.

I went through my standard four-lap warm-up, then blasted down Oops and Poops to prepare for the steeps, then took a lap down All the Way, because I was still in love and, so far this season, the feeling was mutual.

Then I skied Miny by myself, twice. Once to make sure it wasn't dumb luck. The second time to make sure it wasn't a hallucination.

## KIRKWOOD, JANUARY 9, 2021, SEVENTEENTH DAY ON SNOW

Eight inches of new snow on the ground and no more excuses.

Until today, lack of new snow had been my excuse for not jumping off cliffs—or so I kept telling myself. I had a bad history with jumping off cliffs. I have a six-inch metal plate in my wrist as a result of my last attempt to learn how to jump off cliffs. You could say I had cliff-jumping "performance anxiety" if, by "performance," you meant "unable to move" and by "anxiety," you meant "whimpering, blind panic."

Ryan and I decided to settle the issue. We headed for Lightning—a steep bowl that funnels into a series of chutes that drop into a gargantuan gully—with jumpable cliffs from top to bottom.

I was testing out my new Faction Prodigy 4.0s, which are all-black, fat skis, aka deep powder weapons. I brought them out for luck. They each sport the silvery outline of a large white dog, aka a Ghost Dog, aka insert dramatic pause for symbolic emphasis here. . . .

I rode my symbolic skis out to Lightning on our second run. We found zero tracks in the snow. On the first real powder day of the season, we would get first tracks. It was a little pow-pow gift from the universe.

Magic skis plus magic snow plus a highly aesthetic line plus—isn't that the same cliff Ryan jumped last year? The same cliff that I was too scared to jump, the reason why my line list includes the words "Lightning—and jump the damn cliff."

You didn't have to tell me twice.

I sliced toward the cliff and felt my pulse rate spike. Once again, it was much ado about nothing. This wasn't even a cliff. My memory had been shaped by my fear and my memory was flat-out wrong. This wasn't much more than a large boulder.

"Are you sure?" asked the voice in my head.

I jumped off the boulder, stuck the landing, and skied away. "Yup," I said, "I'm sure. And thanks for asking."

## HEAVENLY VALLEY, JANUARY 10, 2021, EIGHTEENTH DAY ON SNOW

A bad night of sleep. A stupid argument. A missed deadline. An unproductive meeting. An out-of-nowhere pile of responsibility. And all before 7:00 a.m.

Sometimes real life invades ski fantasy. Sometimes it piles on so high and so fast that there's nowhere to run and no place to hide. Sometimes skiing feels like the only thing that's keeping me sane.

Today was one of those times.

## KIRKWOOD, JANUARY 13, 2021, NINETEENTH DAY ON SNOW

From the chairlift, Ryan and I noticed that Kirkwood had finally gotten enough new snow to build its first terrain park. A single kiddie line: two beginner boxes into a small jump. "Finally," we said, simultaneously.

This is not how grown men typically respond to kiddie parks.

On any given day, the people who visit Kirkwood's kiddie park are parents with children, ski instructors with students, or the occasional lost tourist. Arguably, I skied Kirkwood's kiddie park more than I skied any other line this season. Arguably, I have no pride left.

The term "kiddie park" is also misleading. The features in the park might be small, but that doesn't mean they can't do some damage.

In Kirkwood's kiddie park, the jump is about three feet high. A rider moving at speed will soar two to three times higher than the height of the jump, then touch down in the middle of the landing, which adds an additional few feet to the fall. Thus, that three-foot ramp was still capable of dropping you ten feet out of the sky—or high enough to hurt should things go wrong.

Ryan and I lapped that kiddie park all morning. My first lap was useless. My second lap was the first time I decided to slide a box. This too did not go as planned. I jumped sideways onto the box, flipped sideways into the air, and slammed sideways into the ground.

Gravity: It's not just a good idea, it's the law.

On my next lap, I was too jittery from my crash and didn't want to risk another slam. So rather than trying to slide across the box, I jumped over it, landing with my skis on the snow and my tails on the box. The impact produced a loud *thwack.*

The *thwack* was proof. Technically, this trick is called a "transition," because I'm landing on the transition between the box and the ground. Technically, a transition is an actual park skiing trick. Technically, that *thwack* was proof that I'd just done an actual park skiing trick in an actual terrain park.

Little wins produce dopamine. God bless dopamine.

On my next lap, I decided to try another transition, but my dopamine-addled brain offered an interesting midair suggestion: "Tail-tap the box." That is, while still midair, twist my body sideways and extend my legs downward—toes up and heels down—until the tails of my skis tapped the top of the box.

I'd never done a tail tap before. Another *thwack.* Another first.

Thanks, dopamine.

My fifth lap was about nose taps. My sixth lap was a blur. So was the rest of the day. Then, on our final lap, I made a mistake that I'd promised myself I wouldn't make.

Several of the neurochemicals that underpin flow are potent pain relievers, meaning they mask fatigue. When in the zone, protecting against injury requires identifying signs of exhaustion before the feeling of exhaustion actually arrives. This was why, on my rule list, I told myself to shut it down if I was underjumping hits or underrotating spins, both of which are clear signs of an empty tank.

Today, I ignored those signs. I was having too much fun. But my body was tired, which meant my body didn't want to try to catch any more air on that final jump, which meant my body decided to initiate the nose butter spin too early and my tails caught and *blammo*—my right shoulder and butt cheek paid the price.

Thanks, dopamine.

### KIRKWOOD, JANUARY 16, 2021, TWENTIETH DAY ON SNOW

Kiko, the one-hundred-and-twenty-pound Maremma sheepdog, loves the drive to Kirkwood. He stands up in the backseat like a surfer on a wave, leaning into the turns, head out the window and the wind in his fur.

Kiko and me—I'm pretty sure we're hunting the same high in the same way.

### KIRKWOOD, JANUARY 20, 2021, TWENTY-FIRST DAY ON SNOW

My new book, *The Art of Impossible,* launched yesterday. All I remember is interview after interview after interview.

Today, I came to Kirkwood on a solo mission. I just wanted the quiet. I just wanted to be alone in the woods. Has anybody seen my Ghost Dog?

### KIRKWOOD, JANUARY 23, 2021, TWENTY-SECOND DAY ON SNOW

Ryan and I got to Kirkwood early. We took one warm-up lap on the lower mountain, then headed up top, dropping into the tight trees beside Rabbit's Run. Three inches of fresh snow on the ground and cold smoke flying off our tails.

On the way down, we noticed that Kirkwood had opened its main terrain park. We saw a pair of medium-size jumps leading into a bigger rail section to the left or a smaller box line to the right. I thought it was going to be okay; I had trained for it to be okay, but seeing Kirkwood's terrain park for the first time—I knew, nothing about this was okay. Yet there was no hiding from the truth. If I was going to learn to park ski, I was going to have to learn to throw tricks off most of those features.

"You ready for this?" asked Ryan.

"No," I said, "but let's do it anyway."

I wasn't ready, but I knew the science and had a plan. The foundational theory that led to this entire Gnar Country experiment: one inch at a time. Start with a well-established motor pattern, add a single motion to that pattern, then practice that new pattern until it's hardwired code. And repeat.

The only hardwired code I had for park jumps were straight airs—just hit the lip, suck up my knees, and try to clear the top of the jump and land on the downslope. Rails were still out of my league. And the only thing I could do on a box was a 50–50—that is, ski straight across the thing.

But a rainbow box caught my attention. It was ten feet long and one foot wide and shaped, as the name suggested, like a rainbow. Could I 50–50 a rainbow box?

Only one way to find out.

I underjumped the first jump, smacking the knuckle and relearning an old lesson: In terrain parks, the correct speed is always a few miles per hour faster than you want to go. I carried that lesson into the second jump, which I managed to clear. This felt like a serious victory and helped keep me calm as I approached the rainbow box.

I skied up the box, I skied down the box, and even I knew that was too easy to count as victory.

"Let's do it again," I said.

This time I cleared both jumps and had enough speed to air onto the rainbow box—which was the first time I'd ever jumped onto a box, rainbow or otherwise.

"Let's do it again," I said.

And again.

And again.

By day's end, I had managed to tail-tap the rainbow box and shifty over a flat box. When the day started, I didn't know how to throw either of these tricks. By day's end, they were both part of my skill set.

I was twenty-two days into my experiment, and just shy of the halfway mark to my stated minimum goal of fifty days on snow. I had managed to learn eight out of the twenty tricks on my trick list. One inch at a time—my plan was actually working.

"Let's do it again."

### KIRKWOOD, JANUARY 27, 2021, TWENTY-THIRD DAY ON SNOW

Want to die young? Talk business with me on a chairlift. Talk flow with me on a chairlift. Talk writing with me on a chairlift. That's all I have to say about that.

### HEAVENLY VALLEY, JANUARY 28, 2021, TWENTY-FOURTH DAY ON SNOW

The Powder Horn Woods always remind me of my time in New Mexico. Old tall trees and no one in sight. Today, the snow was terrible, a mix of hardpack and ice, which also reminded me of my time in New Mexico.

I didn't care. I skied a dozen laps down the exact same line, trying to ski each lap slightly faster than the last. Psychologist Anders Ericsson coined the term "deliberate practice" to describe this kind of repetition with incremental advancement, arguing that this approach to progression is the core foundation of all expertise.

Well, Anders, that's one term for it.

**HEAVENLY VALLEY, JANUARY 31, 2021, TWENTY-FIFTH DAY ON SNOW**

Ryan and I skied twenty-seven laps. Ryan's brother joined us again. He bailed after twelve laps, claiming that work was calling. Ryan doubted his claim. He said his brother bailed because we're fit ski beasts.

He may be right.

**HEAVENLY VALLEY, FEBRUARY 1, 2021, TWENTY-SIXTH DAY ON SNOW**

Skied twelve laps. Still a fit ski beast. Nothing else to report.

**KIRKWOOD, FEBRUARY 3, 2021, TWENTY-SEVENTH DAY ON SNOW**

A huge blizzard blew in overnight. My morning started with twenty-six miles of black ice on the highway. Go too slow and you can't crest the mountain passes; go too fast and you'll sail right off the edge. Sometimes, just getting to Kirkwood can be scarier than anything that happens at Kirkwood.

Not today.

Today, Ryan and I skied twenty-three laps; most didn't go so well. Our fourth run, for example, was a little chute near Dick's Drop. Perfectly untracked powder in there and neither of us bothered to wonder why.

Three turns into the chute, I discovered the problem. My third turn unleashed a mini-avalanche that exposed a medium-size cliff. I had a half second to suck up my legs and straighten out my trajectory before sailing off the edge. While I stuck the landing and skied away, that huge spike in adrenaline did not help the situation.

What situation? Glad you asked.

We'd skied that chute so we could inspect a neighboring line—Pencil Chute—from the bottom. The inspection was crucial.

Of all the lines on my line list, Pencil Chute scared me most. It's an enormous lava tube—nothing more, nothing less. The top is a steep funnel; the bottom requires a seventy-five-foot straight-line exit. The straight line is a five-foot-wide corridor between thirty-foot rock walls. The exit is totally blind, and there are often moguls on the run-out below—thus the cause for inspection.

Ryan and I spotted Pencil Chute last year, when it was early season and the line was still bare rock. Neither of us believed it was skiable. Then we saw it covered in snow, with tracks leading out of it—so someone had skied it. Today, we checked out the bottom, inspecting the conditions on the off chance that maybe, you know, no firm commitment, just a little exploratory curiosity, and see, voice in my head, no need to panic.

Not yet.

The panic arrived next lap. We got to the top of Pencil Chute and realized that a lot of people must try this line, as there was a warning sign up there, reading: DANGER! SEVENTY-FIVE-FOOT STRAIGHT LINE! NO SIDE-STEPPING!

I peered down the funnel, got crazy vertigo, and barely made it out of there without collapsing. Ryan, the fucker, skied Pencil Chute really well. He nailed the straight line, blazed out of the exit, and sent an enormous, forty-foot shifty off a snow drift.

Yet while watching Ryan ski the chute—even though the straight line still scared me to death—I recognized his moves. I could ski this line. If I could find the courage, I could absolutely ski this line.

But not today. Today, my fear levels were too high. I needed to get out of there and calm down. Of course, when you're red-lined, "calm" is a matter of opinion.

To mellow out, we went to the Fingers, and, honestly, what is wrong with me?

We skied Moe for the first time. Are you kidding? It was narrower than Pencil Chute, and just as steep. Four exceptionally unflowy, hop-and-drop turns and then another straight-line exit. It wasn't graceful, but survival felt like half the battle.

Then we decided to try Eeny, which was a mistake. I'd had too many confrontations with mortality. I was overloaded. I was exhausted. Unfortunately, I didn't know any of this yet.

This time, Ryan waited up top and I skied first. The entrance required a five-foot drop onto a sliver of snow beside an enormous blade of rock. As soon as I landed, I got that "skiing through a cave" feeling. It reminded me of Miny, except steeper, narrower, and—holy crap—that's a burly, straight-line exit.

I was too tired and too spooked, and the sight of the straight line pushed me over the edge. I hesitated, just for an instant.

It was enough. My delay slid me onto the rock wall, with my skis heading straight off the side of a cliff. Time slowed to a crawl. I knew I was in trouble, but whatever—I felt calm, cool, and collected.

Ryan felt otherwise. Watching from the top, he did something he'd never done before, not once, not in all our years of skiing together.

He screamed: "HALT!"

So, yeah, good advice. I halted. I realized there was nothing calm, cool, or collected about the situation. My ski tips were dangling over the edge of the rock, about a millisecond away from serious consequences.

But I was more embarrassed than unnerved. I hop-turned a 180, got my skis pointed in the right direction, and straight-lined my ass outta there. Everything in slow motion—like I knew what to do and just did it. Go figure.

By day's end, I counted thirteen moments of pure terror. Apparently,

I was gaining confidence in my ability not to lose my mind at times when losing my mind could mean dying. I guess this was a form of progress, but it sure didn't feel like one.

## KIRKWOOD, FEBRUARY 7, 2021, TWENTY-EIGHTH DAY ON SNOW

The day started with *The Art of Impossible* hitting nine bestseller lists: *New York Times, Wall Street Journal, Los Angeles Times, St. Louis Post-Dispatch, Chicago Tribune,* Amazon, *Publishers Weekly, USA Today,* and the *Southern California Independent.* It was a sweep. I'd never done that before.

I should have felt elated. Three decades of research fill that book. Hitting all those bestseller lists should have felt like a lifetime achievement award. Except now, I was entirely focused on a different kind of achievement. Still, I called my longtime editor Michael Wharton and told him the good news.

"Steven can fly," Michael said afterward.

"Walking on water wasn't built in a day," I replied.* Then I hung up, drove to Kirkwood, and went batshit crazy.

It began with redemption. Ever since Ryan skied Pencil Chute, the gnaw in my gut had been constant. A seventy-five-foot straight line? Could I hold it together for that long? But I knew that if I didn't confront that gnaw, it would chew straight through my season.

We started to go through our four-lap warm-up, but my brain knew about my Pencil Chute plans, and the gnaw was in full effect.

Fuck it—head-on was the only cure.

We skied to the top of Pencil Chute. As we entered the funnel, Ryan glanced back at me, the eternal question on his face: Should I stay or should I go?

---

* This is not my line. Originally it was Jack Kerouac's line. See: Jack Kerouac, *Book of Haikus* (New York: Penguin, 2003), 169.

Staying was bad. Staying meant stopping and stopping meant vertigo. I skidded sideways across the funnel, peering down the chute as I slid. I saw the pattern: the turns, the straight line, the exit. I nodded. Ryan dropped in.

Watching him ski, I realized I had gone two out of three. I recognized the turns and spotted the exit, but that straight line was enormous. Like four times longer than the pattern my brain had predicted.

I let myself drift down another foot. I felt fear start to rise and told myself, under no circumstances was I going to ski this line like a chump. No more shame. No more bad memories. This is Steven 2.0. Now, ski the damn straight line.

I dove into the first turn and knew the drill. We go where we look. The secret to straight lines: Lock your eyes on the exit, lock in your stance, and hang on for dear life. If you crash, aim for the exit, on the off chance you'll crash into open space.

What does zero to fifty feel like on skis? What does zero to fifty feel like when you're hurling down a lava chute, between towering walls of ancient stone, in deep shadows, in your fifth decade on this planet—meaning, definitely old enough to know better?

One Mississippi, two Mississippi, three Mississippi—holy crap, it's taking a lot of damn Mississippis to scream out the other side. The panic arrived in the middle of the second Mississippi. Hello, feeling that I'm going to die. Hello, darkness, my old friend. Eyes on the exit and fuck you too.

I roared as I skied out the exit—then went off and did ten other things that terrified me. Not sure why. Maybe I was hunting my Ghost Dog. Maybe my brain decided to get all the scary shit out of the way in a single day. Maybe it was to counteract all the ego stroke that came from hitting those bestseller lists. One thing's for sure, the mountain doesn't care who you are or what you've done lately. It doesn't care a lick.

In the middle of the day, I led us down a line on the far left side of the Wall, right beside the Cirque, aka the Forbidden Zone. Ryan gave me a look afterward. Apparently, I had been about a half foot from the cliff the whole time. Whatever. The whole day was scary. This was just more scary.

We ended the day with a return trip to the cliff zone between the Notch and All the Way. Once more, I got totally vertigo'd out. Once more, I had to do a kick turn while standing atop a thin spine. This time, I knew what to do: I roared, using the blast of testosterone to jump into the air, spin a 180, stick the landing, and ski the damn straight-line exit.

Just that kind of day.

Yet it was an awful day on the inside. I was exhausted. The flow tap was dry. The weed didn't help. Not high. Not flowing. No grace. Only the occasional sighting of steaze. I suffered vertigo repeatedly. I felt like someone beat me with a pipe.

But on this super high-challenge day, my efforts were completely successful. I skied every line without falling or stopping or choking. Actually, I skied Pencil Chute with style. No choice really, but a definite victory. All told, it was the most physical bravery I'd seen in myself in a single day.

I was fifty-three years old, and how do you like them apples?

### HEAVENLY VALLEY, FEBRUARY 8, 2021, TWENTY-NINTH DAY ON SNOW

I woke up to discover the voice in my head was already awake. "You can't just go around confronting mortality whenever you feel like it," said the voice. "There have been studies. That PTSD stuff is for real. Also, you're out of clean long underwear."

I drove to the mountain anyway. Twelve laps later, the voice finally shut up.

## KIRKWOOD, FEBRUARY 10, 2021, THIRTIETH DAY ON SNOW

The voice had a point: I was freaked out from my first trip down Pencil Chute. My startle response was jacked up to eleven—a sure sign of post-traumatic stress. But with six inches of fresh powder at Kirkwood, perhaps the cure was available.

No warm-up laps, nothing to prepare my body, just the urgent need to clear the residual fear from my system. I led Ryan out past Lightning, into a zone neither of us had skied before. Three turns later, I spotted a five-foot cliff. I wanted the day's first scary act over and done. I skied right off the cliff.

The drop was big. Bigger than expected. "It appears we're still falling," said the voice in my head. "It appears, well, no need to beat around this bush: If this cliff was five feet tall, you would have landed by now."

The voice was right. The cliff was closer to twelve feet tall. But I nailed the landing and skied away and realized my pulse rate hadn't spiked.

"It appears you've improved," said the voice.

Pride goeth before a fall. On our way back to the lift, I popped into a gully and tweaked my back and dropped like a stone.

The pain was crippling.

Flow was my only hope—unless I wanted to spend the next seven days in bed. The neurochemicals that underpin the state decrease pain, fight inflammation, and boost the immune system. If I went straight home, even with saunas, stretching, and such, this could take a week to heal. If I got into flow first, the turnaround time was about three days. How to ski myself into flow when I can barely move—that was the next question.

The answer: weed.

I inched to the bottom of the hill, crawled to my truck, ate a bunch of edibles, took a bunch of aspirin, and waited for everything to kick in.

Once I could move again, we switched into exploratory mode—that is, we switched from trying to use risk and challenge as flow triggers to trying to use novelty and unpredictability as flow triggers.

Exploratory mode took us into Reuter Bowl, at the far edge of the resort. Uncharted territory for us. Yet the map showed single black diamond runs, meaning, I could get dopamine from the novel environment without needing my bad back to do much work.

Exploratory mode delivered. Great snow out there. Even better scenery. Two laps later, I was giddy again.

"Let's go learn some tricks," I said.

"That's the dopamine talking," said Ryan.

"Uh-huh," I agreed. "And it's telling me to get over my fear of throwing 180s off park jumps."

A four-pack of small park jumps, a four-pack of 180s off small park jumps, and my fear of blind lips and backwards landings was gone. Just like that, a little microflow gave way to a little macroflow.

On the next lap, we spotted a small snow berm. Macroflow brings potent pain relief, which explains why I decided to try a switch snow grind on that berm. I slid a 180 up the berm, skidded sideways across of it, then popped a 180 back off. And just like that, another trick on my trick list—now over and done.

Then I noticed a new feature: a six-foot corrugated rail perched atop a small quarterpipe. I realized the same motion I had used on my switch snow grind would slide me up that quarterpipe and along that rail. See the line, ski the line—and, holy crap, did I just grind my first rail?

Once again, pride goeth before a fall.

On my next lap, I attempted my first rail slide on a turbo rail—essentially a long, thick steel sausage. I jumped onto the sausage, flipped upside down, smashed into hard steel, then flopped to the ground.

This definitely didn't help my tweaked back—but that was tomorrow's problem.

Today, we skied a season high of thirty-two laps. Then Ryan led the convoy home, just to make sure my back didn't seize up again mid-mountain pass, because, you know, it would suck to die after making all that progress.

### KIRKWOOD, FEBRUARY 13, 2021, THIRTY-FIRST DAY ON SNOW

A foot of powder fell overnight. The hell drive was intense. The day was odd. The highlight was Ski School Chute—another line on my line list.

I've been scared of Ski School Chute ever since I moved to Tahoe. The cornice was gigantic. The Earth just fell away.

All last season, I avoided the chute entirely. On the first day of this season, I tried to peer over the edge. I got three feet away, was walloped by vertigo, and nearly collapsed. Ever since, I've been creeping closer. Four days ago, I got near enough to get a look at the line. It looked skiable—but only if you were willing to launch fifteen feet off the cornice and miss a big rock on your landing.

Today, Ryan and I saw Ski School Chute living up to its name. We noticed two ski instructors and a handful of ski school students standing at its edge. I'd never seen anyone ski this line before. Screw the vertigo—I had to know how the puzzle was solved.

We started to skate over, but the closer we drew, the twitchier I got. Then I realized that one of the students was a twelve-year-old girl. If she could stand at the edge of that cornice and look down at the line, well, I was out of excuses.

Then one of the instructors dropped into the chute. Turns out, you don't have to leap off the cornice. Instead, you slide in from the side and hop-and-drop to get out of danger, which appeared a significantly easier proposition.

"Son of a bitch," I said aloud, "that's how you ski this thing? I had no idea."

The girl looked over at me. "You can pretty much fall down it and be fine."

So how to nail Ski School Chute? When a twelve-year-old calls you out—what other choice do you have?

## HEAVENLY VALLEY, FEBRUARY 14, 2021, THIRTY-SECOND DAY ON SNOW

I woke up exhausted, but drove to the mountain anyway. I skied sixteen laps. The day ended with a sketchy trip through the trees on Hogsback. Icy, hard-packed snow and not enough of it. Bare rocks. Thick manzanita. Halfway down the line, my ski caught a shrub and sent me somersaulting. I slid to a stop with my face a millimeter from a boulder. Close enough I could kiss the rock.

## KIRKWOOD, FEBRUARY 17, 2021, THIRTY-THIRD DAY ON SNOW

The sketchy conditions continued. Frozen waves of windblown ice and nowhere to hide. Ryan and I started out lower mountain, but gave up after two terrifying runs through the trees. Next, we headed upper mountain. Our goal was the high traverse out to Lightning. If any slope had decent snow, we thought—maybe—Lightning.

Fifty feet later, we were sliding sideways down the hill faster than we could ski across it. After a hundred feet, we gave up and skied groomers to the chairlift. Even the groomers were scary.

"Damn," said Ryan on the lift. "I really wanted to get some work done today."

I understood the issue. Ryan had been following my one-inch-at-a-time approach, building atop automatized movement patterns and using a full complement of flow science to augment his efforts. Just like me, he'd been progressing at a ridiculous clip. The progress was addictive—that was our issue.

"Let's try the park," I suggested.

Lap one was spooky, but Ryan slid a rail and I spun a nose butter and neither of us died. Lap two was better. I pulled off a pair of nose butter 360s and attempted another park jump 180. Sucked, but whatever. Blind landings make for hard tricks, and who needs pride.

On the following lap, Ryan started working on lazy boy slides—laying back on his skis in the same position as someone laying back in a La-Z-Boy recliner, then dragging his hand in the snow to initiate a 180.

The run after that, I copied Ryan's motion and learned a new trick.

A few runs to polish that trick, then we took our game out of the park and onto the hill. And what does group flow look like on a ski mountain?

When Ryan and I started throwing side-by-side nose butters and opposite direction shifties—meaning, he'd hit a jump and twist left and I'd follow and twist right—that's a pretty good sign.

Also, if this skiing thing doesn't work out? Ice Follies, here we come. . . .

### HEAVENLY VALLEY, FEBRUARY 18, 2021, THIRTY-FOURTH DAY ON SNOW

More new snow fell overnight. I went to Heavenly on a solo mission, but met a guy on the chairlift who had driven all the way from Boston for the storm. He said he wanted to ski the gnar.

Well, then . . .

I took him to Mott Canyon, a horseshoe canyon on the Nevada side of the mountain, and home to some of Tahoe's rowdiest lines. A few years back, a ski patroller was found dead at the bottom of one of those lines. It's the real deal.

It was also my first lap in Mott this season, and maybe my fifteenth lap in there total. I didn't know my way around, but I knew Hully Gully—a steep, off-angle tree run with awkward, unpleasant turns. Also, one of the mellower slopes in the canyon.

We slid through the entrance gate and discovered five inches of fresh powder atop solid hardpack with only occasional bare patches. But the new snow was fantastic, which wasn't always the case in Mott, and then it was nonexistent, which was often the case in Mott.

Mott's something of a sucker's bet. If you pull into the canyon for the first time, it appears the steepest part of the run is at the top. But that's an optical illusion. The canyon is actually convex and there's a second crux in the middle: essentially, a long cliff band that's often too steep to hold snow.

The good snow up top had lured us into a false sense of security. After the smoothest run down upper Hully Gully I'd ever skied, we sliced deeper into the canyon, blazed into the second crux, and discovered that all the new snow had already been skied off. What remained was boilerplate ice and big boulders. We both did the rock dance to survive.

"Spicy," said the skier from Boston afterward.

"Wanna try extra-spicy?" I asked.

Extra-spicy was Bill's, which I'd heard about but never skied before. The snow was better in there, the terrain fiercer. I knew, if I knew my way around, there'd be a flowy line down Bill's, but I didn't find it today.

"Extra-spicy," said the skier from Boston.

"Wanna try nuclear?"

Nuclear was the Y, which was one of the gnarlier lines in the Canyon. There's a fifty-degree section up top and an even bigger cliff below. Steep as hell and a true "no-fall" zone.

I pulled onto the steeper section up top and got hit with instant vertigo. I dropped into my first turn before the spinning became a problem. I skied like a dickhead, sure, but pulled it together at the bottom and finished strong.

"Nuclear," said the skier from Boston.

"Wanna try thermonuclear?"

"Nope," said Boston. "I'm gonna try alcohol."

I knew how he felt, but I wasn't done. We parted ways. I went exploring. I did not go for thermonuclear. Instead, I went in search of cool down laps.

I found two different tree lines with high flow potential, skiing five laps down each at silly speeds. What can I say? After scaring myself in Mott Canyon, doing the gravity dance through tight trees felt downright relaxing.

## KIRKWOOD, FEBRUARY 21, 2021, THIRTY-FIFTH DAY ON SNOW

Gray skies and tricky conditions. Six inches of powder on top, thick sheets of ice below. Ryan and I warmed up and got on the Wall. Norm's Nose was our destination, but the traverse was boilerplate. My fat skis couldn't find an edge. I tried five turns, none held, and nothing good comes from this much sideways.

We vetoed the Nose, pointed our skis downhill, and got the hell outta there.

But, I knew, the ice wasn't my problem. My problem was Pencil Chute. Even though my first trip was a success, the chute scared me. I was twitchy. The twitch was a sign of too much norepinephrine. The neurochemical increases excitement and improves focus at lower levels, but get above a certain threshold and excitement becomes anxiety, focus becomes hypervigilance, and neither is good for the soul.

To reset my nervous system, I needed another shot at Pencil Chute. Yet the second time through was always worse than the first, as this time I knew what was coming.

Of course, I sketched out on the approach and was too nervous to try the chute. Of course, I skied to the bottom and watched Ryan nail the line. Of course, he claimed the snow was great in there—like better than anyplace else on the mountain and so much better than it had been on our last trip through.

I tried to believe him. I knew that I had to ski Pencil Chute today or live with the fear. I was tired of living with the fear.

We got back on the chairlift. There was a U2 line stuck in my head: "Then I floated out of there." I suspected that this was what death would feel like. And I floated outta there. . . . Like: *see ya, thanks,* then gone.

Was it a premonition? It was something.

This time, we got to the top of the funnel and Ryan didn't hesitate. He charged into the line, made the hard left turn, pinned it through the rock walls, and vanished from view. I was alone. I was staring down a seventy-five-foot straight line. If I stared for long, vertigo would set in. I slid down a foot, then another, but side-slipping Pencil Chute—no way that counted as victory.

Fear of future shame launched me into the chute. I nailed the first turn, nailed the second, spotted the exit, locked my eyes, locked my stance, started to accelerate—and, what the hell?

It felt like my right ski was riding on marbles. I glanced down and caught a fast glimmer. Icicles had fallen from trees and blown into the chute and my right ski was riding atop a dozen of them.

And then it wasn't.

From below, Ryan saw my skis start to wedge and my body start to tip forward. Apparently, I was about to somersault down Pencil Chute.

I felt the deep calm that always precedes a bad crash. I knew this feeling. I knew what it meant.

Not this time. This time, I discovered another gear. Later, I dubbed it: "Fuck this! Not dying!"

It was a moment of berserker rage. A wild surge of testosterone and adrenaline. I battled against my brain, battled against gravity, got my torso upright, got my skis tracking, blazed through the straight line, and I floated outta there. . . .

In the end, we did eighteen laps. The highlight was a tree line off Chair 5. Jump past the stump, cut hard right, and find the entrance tucked between trees. Tight turns in there, and all with consequences. Perfect for fast geometry.

Then we made it pretty: nose butter 360 off the initial hit, shifty off the second, then the stump jump, a hard right through the gap in the forest, and, finally, a little 180 off a side hit at the bottom.

On our last lap, as I bounced off the last jump, I caught sight of a dad and his kids in the middle of the landing. Dad was tucked behind a pine tree; the kids were plopped down on the snow. I was already starting to spin, but knew if I tried a 180, I was going to land backwards and maybe without enough edge control to miss the kids.

Midflight change of plan.

I felt bad about it afterward, but threw a huge nose butter 360. I landed directly between the kids, facing forward and in full control. I skied away like nothing had happened, but not before I saw Dad's jaw drop open.

But the rules are the rules. If I do a trick, Ryan has to do the same. In other words, because Ryan was behind me and didn't yet know about Dad and the kids, and because I had just thrown a nose butter 360, Dad's jaw stayed dropped open because Ryan did the same damn thing after me.

The kids, meanwhile, applauded.

### HEAVENLY VALLEY, FEBRUARY 22, 2021, THIRTY-SIXTH DAY ON THE SNOW

The weather report: sunny skies and heavy crowds.

In the lift line, I bumped into my old friend Sofia Mileti, a ripper I knew from New Mexico. Sofia was in Tahoe for the winter. She'd been hiding from the crowds in Mott Canyon. We teamed up for a few laps.

Sofia was warmed up and already skiing at full Mach. I tried to stay on her tails. She led me on a zigzag chase, stitching together sections of Hully Gully, Widowmaker, and Snake Eyes. We ended on a line of hero bumps at the bottom of Ernie's. Sofia liked the hero bumps because they reminded her of her childhood on the Park City freestyle team. I liked them because they were in full view of the chairlift, and hero bumps are called hero bumps because they make everyone look good.

I carried that confidence back to the Y. Sofia had never skied it before. I uttered those famous last words: "Just follow me."

Unfortunately, I wasn't warmed up. I skied the top of the Y like an idiot. It got worse in the middle. The crux was sheet ice. Twenty turns of the don't-die variety. Pure survival skiing. Everyone on the run was skiing for their lives and looked crappy doing it, but I took my shitty skiing personally.

In an effort to shake off my shitty skiing, I ate an edible and tried to forget. I forgot all right. I forgot to read the label. I thought it was a ten-milligram cookie. It was actually a hundred milligrams. An hour later, somewhere in the Powder Bowl Woods, the trees started talking to me.

### HEAVENLY VALLEY, FEBRUARY 23, 2021, THIRTY-SEVENTH DAY ON SNOW

I took the afternoon shift and did twelve laps. Sofia was around, but she had taken a tree branch to the face in Milky Way Bowl and was pretty shaken up. We tried to keep our adventures mellow.

A few laps later, we abandoned the mellow. It wasn't intentional—it was neurochemical.

We were skiing Express Line, a tight maze of moguls and rock drops that sits directly below a chairlift. Having an audience raised the challenge level. Nailing our performance rewarded us with dopamine—aka, the neurochemical reason that show-offs like to show off.

Of course, we came back for an encore.

Afterward, Sofia said she was done for the day. I felt the same, but took a final spin through the Powder Bowl Woods and apologized to the trees for anything inappropriate I might have said while stoned. Then, I went home.

## HEAVENLY VALLEY, FEBRUARY 25, 2021, THIRTY-EIGHTH DAY ON SNOW

The conditions were nasty, and Kirkwood was worse than Heavenly—so we went to Heavenly. It was nasty, all right. Rock-hard snow and nothing to do but high-speed groomers until the sun warmed things up.

Yet the conditions were the least of my problems. I was stuck in gear limbo. I needed my K2 Mindbender 108s to hold an edge on ice, but they're not designed to ski backwards and can be tricky to land switch. My Prodigy 4.0s were fantastic at skiing backwards, but they were too fat to hold an edge on the ice and that made them deadly in the steeps. And my Faction CTs were useless. Maybe it was the mount point, maybe it was the ski design, maybe it was my shitty skiing—but every time I tried to ride the CTs, they tried to murder me.

I was tired of fighting my gear, but skied the fight right out of me. By run ten, the voice in my head was done complaining. On eleven, Ryan found some good snow in the trees. We celebrated by skiing ten laps down the same line, jumping off everything in sight, and then, on the next lap, trying to jump even farther.

On our last lap, Ryan went flying into the forest, cut left between trees, leapt off a boulder, and made a tiny, midair head twitch. I saw the twitch just as I hit the boulder.

"Oh crap," said the voice in my head.

Buried in my subconscious was an entire file of Ryan's nonverbal lexicon. I knew all his twitches. I knew that this particular head twitch meant that there was zero snow on the other side of that boulder. I knew I was in big trouble.

Yup, midflight big trouble head-twitch suspicion confirmed: grass, rocks, and trees below. Not a snowflake in sight.

But Ryan saw something I didn't. He arced sideways midflight, landed on a tiny crescent of white, then leapt again. He sucked up his knees and soared toward a second snow patch about ten feet away. High-speed survival skiing. Less fast geometry and more drunken pinball. But when it was over, it made me laugh, in that lucky-to-be-alive kind of way.

### KIRKWOOD, FEBRUARY 27, 2021, THIRTY-NINTH DAY ON SNOW

Serious ice. Serious cold. Like, ten degrees all day, ten below with the wind chill. We did a few warm-up laps, but they didn't help. We knew the cold would hamper our reaction times, tilting us out of the challenge-skills sweet spot and making flow a nonstarter. Exploratory mode was our only hope.

We decided to check out the mountain's sunny side. The sun was out, but so was the wind. It was ten degrees colder with the wind. The lie of the sunny side.

Next, we traversed in the other direction, investigating the far edge of the Palisades. Neither of us had been out there before. We discovered five-hundred-year-old western junipers. Golden orange bark striping wildly twisted trunks, a tree built to last a thousand years.

There were a few inches of fresh snow hidden beneath those trees, and boilerplate ice everyplace else. Whatever, it felt nice to float.

In the afternoon, we decided to use the terrible conditions to our advantage. Could we play fast geometry down Oops with the chute in seriously bad shape?

Interesting question.

The top of the chute was frozen moguls; the middle was just frozen. One of those runs when you really needed to trust your edges. Mine held. And doing the gravity dance down sheet ice produced instant flow.

Admittedly, there's no such thing as "instant flow." If I was keeping office hours, I would say that this experience was packed with flow triggers—the psychological preconditions that lead to neurobiological reactions that produce both individual flow (me, alone, in the state) and group flow (me and Ryan, together, in the shared version of the state).

On the individual side, there was *risk*, *novelty*, and *unpredictability* in the terrain. To ski this terrain, I had to make *creative* and *embodied* decisions at the very edge of my *challenge-skills sweet spot*. That's six triggers for individual flow.

On the group flow side, the list is longer. Before we get to that list, one detail is crucial: Flow follows focus. The state only arises when we focus all our attention on the here and now. Thus, all of flow's triggers amplify attention. The difference between individual flow triggers and group flow triggers? The group triggers drive a shared version of attention, where two or more people become absorbed in the same experience and, often, in the same details of that experience.

More specifically, when Ryan and I skied that icy line together, to create group flow, we relied on *shared risk, shared goals, good communication, equal participation,* and *close listening.* Again, if I was keeping office hours, I would point out that these same group flow triggers can be used in a business meeting, or when you're trying to talk to your kids, or any other place where two or more people get together.

Of course, since I'm not keeping office hours, forget I mentioned it.

One thing's for sure: All that flow got the residual trauma from my second lap down Pencil Chute out of my system. Only took a high-speed day on super bitchy ice to clear my head.

## HEAVENLY VALLEY, FEBRUARY 28, 2021, FORTIETH DAY ON SNOW

The mountain was packed. I decided to hide from the crowds in Killebrew Canyon. I was alone, which made this a dicey decision. Killebrew is home to some uncomfortably rowdy runs, including Pipeline, the steepest run at Heavenly—or so I'd heard.

I had never skied Pipeline before. In fact, I'd only been to Killebrew on one other occasion, wherein I got quickly lost, missed the final turn, and ended up wandering through the trees, far out of bounds. I scared myself pretty good. Eventually, I found my way to the Rim Trail for that long hike back to the resort—swearing the entire time.

This time, the same thing happened. Once again, I was alone. Once again, I was lost. Were there cliffs below me? I had no idea. Was there even snow below me? I had no idea. My heart thumped. My vision clouded. I did not like where this was heading.

My worst fear in skiing is being alone, lost, maybe above exposed cliffs, with the threat of vertigo looming. Before that panic could take hold, I dove into my first turn and skied the line like I meant it.

But the line wasn't actually that steep. And there were no cliffs below me. Five turns in, my brain downgraded the threat level. Suddenly, my vision cleared and I realized the canyon was gorgeous and the turns were flowy. I started smiling. I got the sense that Killebrew could become my favorite spot at Heavenly, but not today.

Today, I missed the turnoff again, skied to the bottom of the canyon, and had to hike that damn Rim Trail back to the resort. I spent most of that trek thinking about how many times I've been lost and scared in the mountains. Too many times.

I skied six laps in Mott Canyon afterward, hunting progress. I wanted one flowy line. I tried to find it on Hully Gully, Widowmaker, Snake Eyes, Bill's, Ernie's, and finally the Y. It didn't work. My Ghost Dog was nowhere in sight.

### KIRKWOOD, MARCH 1, 2021, FORTY-FIRST DAY ON SNOW

Cooked from yesterday—or so sayeth my muscles. The ache tried to talk me out of skiing. The ache lost that battle. I'd been wanting a solo mission to Kirkwood and decided to take it slow and see what happened.

It was another lesson in how much my body lies about fitness readiness. I woke up unable to move. I started to move. I moved my ass to

Kirkwood—and did twenty laps. Twelve were nonstop burners down Oops, which was still sheet ice, which was also the reason I wanted a solo mission.

I needed the ice. I was testing a theory.

Stanford neuroscientist Andrew Huberman once told me that you can fight fear with peripheral vision. When we're looking at the world through the corners of our eyes, the brain thinks: *OK, all is chill, you're just checking out the scenery.* This activates the parasympathetic nervous system, aka rest-and-relax mode.

Tightly focused vision produces the opposite reaction. It activates the sympathetic nervous system. Instead of rest and relax, we're kicked into fight or flight. Once there, our vison contracts further, so we can stay laser targeted on the survival threat directly in front of us.

Today's experiment built off this idea. It was sparked by a comment made by Ryan. Unless we're in tight trees, which are my specialty, Ryan's a slightly faster skier than me. A few weeks back, he said that keeping his vision pegged far down the run was one of the main secrets to his speed.

"It allows me to see more of the hill," was how he explained it.

"Seeing more of the hill" sounded like peripheral vision. So I started to wonder if keeping vision pegged farther down the run could trigger parasympathetic activation. This would lower fear levels, expand the challenge-skills sweet spot, and perhaps allow me to find a backdoor entrance into flow.

That was why I skied twelve laps down Oops in boilerplate conditions. The crux of the run resembled a sideways skating rink. It was the kind of terrifying section that demanded laser focus. But what would happen if I kept my laser focus farther down the hill? If this counted as peripheral vision, I should calm down and see more of the slope. If it didn't, I would probably crash, and that would be proof of a different kind.

I didn't crash. In fact, by keeping my vision farther down the hill,

I saw more line possibilities, and this made the run feel slower despite the fact that I was skiing faster. This shift also lowered fear levels, which was further proof of parasympathetic activation.

Once again, the biology worked as advertised.

### NORTHSTAR, MARCH 3, 2021, FORTY-SECOND DAY ON SNOW

Maybe my favorite day of the season and it was spent lapping Northstar's kiddie park. With the threat level low and my confidence high, as long as I kept my ego out of the equation—the fact that I was, after all, lapping a kiddie park—the opportunities for advancement were considerable.

The results: My first zero spin—that is, take off backwards, land backwards, and throw zero spins along the way. My first "tweaked" safety grab, grabbing my ski beneath my boot and tweaking it—aka yanking it—sideways. Dozens of nose butter 360s. And definitely the best 180 I'd ever thrown off a park jump.

This brings us to a small digression on mindset.

Scientists define mindset as "core assumptions that orient us to a particular set of expectations, explanations, and goals," to paraphrase Stanford psychologist Alia Crum. These assumptions play a crucial role in our later years. Research dating back to the 1970s shows that a positive mindset toward aging—that is, viewing the second half of our lives as a time of exciting possibility—prevents cognitive decline, preserves physical prowess, and extends longevity. The most famous example of this impact comes from the twenty-year-long, thousand-participant-thick Ohio Longitudinal Study of Aging and Retirement, where researchers found a positive mindset toward aging produced a staggering extra seven and a half years of life.

For certain, my first forty-two days on snow had shifted my mindset toward aging. This was not the by-product of effort. I didn't try to replace my older, limited view of my later years. Rather, when confronted with the evidence of nose butters, zero spins, and tweaked safety grabs, that older view didn't stand a chance.

To close out the afternoon, I tried my first 50–50 on a turbo rail. Ryan tried it too. Our improved mindsets did not help. Both of us skied onto the rail with our skis held together. In both cases, our skis slipped apart, and we smashed, testicle first, into the steel.

Alas, there's always tomorrow. . . .

### NORTHSTAR, MARCH 4, 2021, FORTY-THIRD DAY ON SNOW

Today marked my forty-third day on snow. This meant I'd tied my lifetime season high for ski days. It should have been a major milestone. It didn't feel like one.

I woke up bone-tired and anxiety overloaded. The tiredness was expected, as it was the by-product of pushing my body so hard. The anxiety was surprising, as it appeared to be the by-product of pushing my mind so hard—that is, the by-product of success.

I'd been ticking off lines and tricks at an astounding rate, at least for me. I'd never learned anything this quickly before—and that was the problem.

Showing up at the hill, day after day, knowing I was there to do something that scared me; this had become tricky after three straight months. There was a cumulative psychological weight to confronting fear, even if those confrontations ended in success. I hadn't expected the weight, as I'd never made this much progress before.

Another thing I hadn't expected: the addictiveness of progress. Sure, with all the time I've spent studying peak performance, I probably should have seen this one coming. Then again, does anyone see addiction coming?

Either way, I ventured into Gnar Country with low expectations and high curiosity. I got my ass kicked the first month of the season—torn rotator cuff, right hip bruise, right ass bruise, etc.—which met my low expectations and dampened my curiosity. I turned a corner after being forced to jump the cliff on Chamoix, which opened the door to larger cliff jumps and led to me ticking off more of the lines on my line list.

Those big lines were so scary, they actually made the terrain park feel safer. This lowered my anxiety levels and enhanced my performance and led to a string of successes, including the two new tricks I'd learned yesterday. So while the cumulative weight of all that fear was beginning to fray my nerves, I was making so much addictive progress—no way was I gonna stop.

Ryan felt the same.

Thus, while today should have been a rest day, neither of us could resist returning to Northstar for another round of jump-off-everything madness.

The excitement kept us going for a while. It sent Ryan on a rail-sliding tear. It made me forget the ache in my body. Then, by the end of the day, the ache became a pain. Then I remembered. All the way home, I remembered.

# Chapter 6

**HEAVENLY VALLEY, MARCH 6, 2021, FORTY-FOURTH DAY ON SNOW**

I did the math.

Today was my forty-fourth day on snow. I was six days away from my preseason target of fifty days. I didn't want to jinx my progress, but the math suggested I could ski more than fifty days. So how many days *could* I ski?

Good question.

I started my season strong and felt even stronger now. Desire had not waned. My hunger for progress was enormous, which was fueled by all the progress I'd already made.

All this progress resulted from having my five major intrinsic motivators (curiosity, passion, purpose, autonomy, mastery) and three tiers of goals (mission-level, high and hard, clear and daily) all pointed in the same direction. On the motivational side, my *curiosity* toward park skiing fed my overall *passion* for the sport, which fed my larger "decode peak performance" *purpose* in life. In turn, these motivators were fueled by *mastery*, or my steady park skiing progress, and *autonomy*, or my freedom to pursue mastery—that is, my ability to ski when I wanted to ski. Of course, my ability to ski when I wanted to ski resulted from adhering to my ridiculous—3:00 a.m. to 8:00 p.m.—schedule, which was the reason I needed all this motivation in the first place.

On the goal side, I had daily clear goals (ski sixteen laps) that fed my high-hard goal (learn to park ski) that fed my mission-level goal (advance the science of peak performance). For these reasons, I needed to be very careful with my answer. I was about to create a new goal. I didn't want to mess with my motivational stack, not when it was working so well. Plus, the "no more shame" clause in my contract was clear: If I set a physical goal, if I said it aloud, I would have to accomplish it.

I considered a hundred days on the snow as my new target, but didn't want to set myself up for failure. A hundred days would allow no time off. It meant I couldn't get injured. It meant skiing into June—which was only possible if the resorts stayed open that long or if I committed to a lengthy Mount Hood mission over the summer.

A hundred days felt dicey. Would the pressure to always be skiing hurt my motivation to keep skiing? Would this pressure tip me out of the challenge-skills sweet spot and interfere with my progress? Too dicey.

I settled on eighty-six days on the snow. My all-time season high was forty-three days, so eighty-six would double my best. This seemed to check the requisite motivational boxes. It had the right amount of symbolic weight.

Later, I would realize that my desire for park skiing progression was at odds with my desire to ski as many days as possible. Park skiing was about power and agility. It meant explosive, high-energy days. Skiing as many days as possible was about endurance. It meant conserving energy along the way because the way was long. Plus, park skiing required more recovery time than I had anticipated and, because of my endurance goal, definitely more recovery time than I was allotting.

But those were all tomorrow's parties.

Today, I skied twenty-seven laps. I didn't intend to ski that much, but twelve laps in, I dropped off a small boulder, got that feeling of flight, and flow was the result. "Never waste a flow state" is another rule, so I skied fifteen more laps, trying to push myself slightly harder on every one.

When I was done, I drove home, thinking about my new goal: eighty-six days on the snow. I had said it aloud. I had called it a goal.

And what was I thinking?

### HEAVENLY VALLEY MARCH 8, 2021, FORTY-FIFTH DAY ON SNOW

Afternoon mission. One inch of new snow on the ground. It was just enough fluff to get into trouble, but not enough fluff to get out again.

I got into trouble.

The snow lured me into speed, the speed lured me into air, the one inch of snow didn't provide much of a landing—you get the problem.

First, I was skiing through the trees, whipped around a tall pine, and blindsided a wind lip. The lip formed a giant ramp that launched me into the air. I flew past a pine tree and the voice in my head asked a familiar question: "I wonder if we're going to die now?"

I didn't die, but it was a close call. One inch of new snow meant the coverage was thick enough to hide what lay beneath, but not thick enough that I would glide over it. This explains why, on my next run, in the middle of a hard carve, my tails caught a tiny stump and I was again launched into the air. This time, I managed to dive headfirst into a snowdrift before being flung into a tree.

Two runs later, I was trying a nose butter when my tails caught again. That one thumped me hard. Those two close calls and one hard thump were enough for me. I got off the hill and went home, feeling lucky to be intact.

What I didn't know: My luck wouldn't last.

## KIRKWOOD, MARCH 9, 2021, FORTY-SIXTH DAY ON SNOW

Strange weather. The Sierras were bone dry, but a massive blizzard had been pounding Kirkwood for days.

Icy roads. Whiteout conditions. Over the Carson Pass, the pickup in front of me got a little rambunctious on the gas, and its back end cut loose. The truck fishtailed across the road, spun in a circle, the driver wide-eyed through the window as I swerved around him. Later, I realized this had been a life-threatening situation. At the time, it was a slow-motion ballet with the sound turned off.

But the hell drive paid dividends.

Ryan and I got first chair. The snow was so deep we had to point them straight down the hill. All we could do was lean back and laugh. The powder was bouncing off our thighs. It was ski porn powder—the lightest and deepest either of us had ever encountered.

On our third lap, I decided to use the new snow to my advantage. I wanted to settle a grudge with Main Nostril, a straight-line chute down the center of Norm's Nose, maybe fifty feet long and five feet wide.

Last year, I'd attempted Main Nostril on two separate occasions. Both times, I got gripped at the top of the straight line and had to sidestep to the bottom, hating myself for every side step.

Today, I skied into the chute, went for a turn high on the left wall, and triggered a mini-avalanche. It swept out my feet, flipped me upside down, and buried me up to my waist. I was now at a dead stop, feet pinned above my head, dangling above the crux of Main Nostril. It was the same spot where I'd sketched out last year. Twice.

To dig myself out, I had to do an upside-down sit-up, then pole-whack the snow by my feet. Ten whacks later, my skis popped loose—only now I was sliding, upside down and backwards, down the chute.

I did a twisting somersault to get my skis beneath me, then held it together as I shot out the exit. No way that counted as victory. Ryan didn't even bother to ask. He knew we were coming back to ski this line again, immediately.

The second time, I just blasted into the chute and blazed out the exit. Too easy. I'd settled that grudge without any fanfare. But I had an idea: The blade of rock that made up Main Nostril's left wall had an additional chute running down its center. It looked skiable.

Famous last words.

We realized our mistake as soon as we got back to the top. The chute was actually a cliff, with just a light dusting of snow.

"I think I see a line," said Ryan.

"Nope."

"I think I'm gonna give it a go."

It went, all right. Ryan's first turn triggered another mini-avalanche. If he didn't stay ahead of the slide, he was going to be skiing over bare rock. The only solution was to suck up his knees and sail off the cliff.

But as Ryan tried to suck up his knees, his right ski caught raw stone and ripped him sideways. A second before he smashed into rock, Ryan managed to tear the ski free and clear the cliff.

"Fuck me," I said, as he sailed out of danger.

I was not out of danger. I was still trapped. The slide had wiped the snow from the rocks. My only choice was to ski to my right and pray for a possible side entrance into Main Nostril.

But there was no side entrance. There was just the big blade of stone. My final option was to jump a ten-foot cliff, twist in the air to center myself between the rock walls, and stomp the landing or else.

There was no time to find a better solution. As soon as I'd skied onto the cliff, I'd triggered a mini-avalanche of my own. It was the same situation that Ryan had encountered, and it had the same solution: either beat the snow down the cliff or ski over bare rock.

I jumped into the air, turned sideways, cleared the cliff, stuck the landing, and shot into the straight line. Far below, I saw Ryan start to raise his arms in victory. Then, I saw his eyes go wide in terror and heard a voice in a place no voice should be.

At the same moment I jumped off that cliff, a six-foot-one, two-hundred-and-ten-pound gentleman skied into the Main Nostril, accelerated into the straight line, and reached terminal velocity a millisecond before he reached me.

T-boned—that's the technical term for what happened next. My experience was a sharp pain in my shoulder and the sensation of flying backwards through the air. I was bear-hugging the other skier. Then we started to flip. I could see the chute's right wall inches from my face. I could see the pockmarks in the lava rock. I knew the odds that two entangled skiers could somersault down Main Nostril without bashing into lava rock would be like winning the lottery and getting struck by lightning in the same afternoon.

"Those are not good odds," said the voice in my head.

Impact. Then silence. Then the realization that I had stopped moving and couldn't breathe. I was buried under snow. I was buried under human. I couldn't see. I found my face. I got the snow wiped off my goggles.

The first thing I saw was the other skier's eyes. We had landed inches apart. Immediately, I knew, he was fine. There was no pain in his eyes. My eyes must have told a different story.

"Are you okay?" I heard him say. "Do you know your name?"

"Steven," I said, sitting up and testing limbs. "My shoulder's jacked, but I think I can ski. You're okay?"

"I'm okay."

Then, he was gone.

Later, I would wonder about his speedy departure. Just then, I was in crisis mode. I knew I had about a minute of adrenaline left before the pain arrived, and more skiing might not be possible. I jumped to my feet and shot down the trail, shouting at Ryan as I passed: "Have to get someplace the ski patrol can get to me."

Bent over and cradling my arm, I made it off Norm's Nose and back onto a groomer. The pain was still manageable. My next thought: Get off the hill.

I don't remember what happened next. I do remember being back at my truck.

"Hospital?" Ryan asked.

I checked for dire injuries. I wasn't bleeding. No bones were sticking out. All of my joints, save my right shoulder, still worked.

"Not yet," I said.

I took three Advil, slathered my shoulder in CBD ointment, and, as blood sugar crashes after an injury, ate a bar of chocolate. Then I smoked a joint.

Marijuana decreases inflammation, deadens pain, and blocks the establishment of long-term fear memories. Research shows it can stave off PTSD. I wasn't taking any chances. I smoked another joint.

The pain started to subside. I decided not to go to the hospital, or not yet. What I needed was to ski again. If I could ski myself into flow, the neurochemicals that underpin the state would provide more anti-inflammatory pain relief, and further protection against PTSD.

Test lap on Chair 5. I had no use of my right arm. I couldn't pole plant, which made it hard to turn. Instead, I was forced to straight-line the powder and lift up my tips, doing tail drags to slow down. But tail drags are fun and one lap begat two, which begat three.

"Let's investigate the trees," I said on lap four.

Powder skiing through tall pines and hello, Ghost Dog. Then the pain was gone. Then we found a few miracle lines—that is, lines that were only skiable in perfect conditions. In the middle of those miracle lines, apparently, I jumped a cliff on Oops that I'd been terrified to jump last year. It was a line on my line list and I didn't even notice. Ryan pointed it out some four hours later, when we were already driving home.

Once I got home, the pain returned. My wife had to help me undress. I screamed when she pulled my sweatshirt over my head.

## KIRKWOOD, MARCH 13, 2021, FORTY-SEVENTH DAY ON SNOW

When I woke up, my right shoulder was the size of a cantaloupe. There was a baseball-size bruise on my neck, a bowling ball on my chest, and something that looked like a map of Massachusetts in the middle of my thigh. My inner thigh was dotted with broken capillaries, like I got prison inked by an octopus. Also, my brain was not pleased. Flashbacks, night sweats, and my startle response amped up to the max—the PTSD had arrived.

I should go to the doctor. Instead, I went to Kirkwood. There's a saying in action sports: "Most people call it trauma—we call it Monday." In other words, I was going to reach my initial goal of fifty days on snow, even if it killed me. I had three days left.

"Suboptimal," said the voice in my head.

"Copy that," I replied.

Today's goal was a handful of baby laps. Groomers. Nothing heavy. I made my position clear to Ryan in the parking lot—under no circumstances was I getting on the Wall.

The issue with my position was crowds. It was some kind of holiday weekend and Jerry was in town. The only lift without the crowds was the Wall. Thus, on our second lap, I got on the Wall.

"You should have taken more time off," said the voice in my head.

"You should mind your own business," I replied.

We skied the Wall seven times. The first three laps were awful. Then I spotted a new line into the woods and decided to go exploring. Two turns in, what I thought was a small cluster of trees turned into an entire section of mountain that I didn't know existed. Exploratory mode—as usual—delivered.

Ryan and I skied three more laps through our new woods line, each one better than the last. Sure, I was injured, but I didn't notice the pain, and that was the point. Action sports, contact sports, and old age—three activities where learning to use flow to play through injury is a nonnegotiable.

And once you're in flow, the euphoria, the deep sense of meaning, the turbo boost to purpose, that's the real point. Injuries heal. An increased sense of meaning and purpose—that lasts a lifetime.

## KIRKWOOD, MARCH 17, 2021, FORTY-EIGHTH DAY ON SNOW

Over the next few days, I got worse instead of better. I stepped up my recovery game. I got a massage. I went to a chiropractor. I doubled down on my sauna time. I doubled down on my yoga. I could feel scar tissue amassing in my shoulder and back. I needed the chiropractor and the sauna to loosen up my body and the massage and the yoga to bust up the scar tissue. Would it work? Could I ski?

Ryan and I went to Kirkwood to find out.

I began by running experiments, testing my back, testing my shoulder. Could I turn left? Could I turn right? Could I take the impact of bump skiing? Could I handle airtime and landings?

The pain was tolerable, that was the good news. Neither of us could get into our bodies, that was the bad news. We turned to our go-to run in times of doubt—a little slopestyle line through the woods that ended at the top of the kiddie park.

Ryan took point. The first couple of features were perfect for snow grinds, and those are perfect for a guy with no use of his right arm. Then Ryan threw a nose butter 360 off a side hit and, according to the rules, I had to do the same.

I was too tentative. Halfway through the spin, I ran out of juice and started falling toward the ground—my damaged shoulder leading the way.

"Kick," said the voice in my head.

"What?"

But there was no time to argue. I kicked. My feet shot straight out in front of me and plopped me, safely, onto my ass. It was pure slapstick. Ryan and I started laughing.

The science of embodied cognition is built around a simple idea—we don't think just with our brain; we also think with our body. And, as this bit of slapstick reveals, often times, our body knows more than our brain. Yet learning to listen to those signals—that can take some practice.

I took three more falls over the next few hours, none as funny as the first. I landed on my right shoulder three times in a row. Each time hurt. But it was just pain. As far as I could tell, I hadn't done any additional damage.

The additional damage came after lunch. We were diving into a woods trail between Oops and Olympic. What I didn't notice was the branch jutting across the run. It caught my left biceps—the force nearly yanked me off my feet.

I roared as I ripped my arm free. Not the good kind of roar. The pain pissed me off. I was tired of the pain.

"Fuck you, pain!" I shouted, skiing faster.

"Fuck you, pain?" Ryan asked, as I blazed by.

But the anger did its job. The faster skiing, as well. We raced down the mountain and into the terrain park and there, my Ghost Dog was waiting. Game on. Or, more specifically: Finite game off. Infinite game on.

## HEAVENLY VALLEY, MARCH 18, 2021, FORTY-NINTH DAY ON SNOW

It's a miracle—I can use my arm again. Another miracle: It was the first day of spring conditions. A test lap confirmed both miracles: My shoulder worked again, and the snow was soft and grippy.

Ryan and I warmed up with superfast laps down the trees on East Face. So slushy we could point 'em straight through the moguls. Then we took Canyon lift to the top and started throwing nose butter 360s off everything in sight. We were both getting great pop off the jumps, which means my shoulder pain must really be gone. When injured, there's an unconscious governor that regulates everything from perception (how steep a line appears) to pop (how much amplitude I get off a jump). As we were bouncing around the hill, the governor must have been off duty.

It had been a little while since Ryan had been to Heavenly, so I showed him my recent discoveries. The highlight was a little double stager—that is, two jumps positioned directly in a row, so the landing from the first jump becomes the takeoff for the next. The double stager provided more proof my shoulder wasn't actually broken, as the governor doesn't allow such things when on the job.

We ended with a lap through Ski Ways Glades. There was great snow everywhere. We were in high-speed drift mode—legs loose, slarving wide turns, sucking up knees over the bumps, hunting the right line, then *bam*—unleash the fury.

At least, that's how it felt. Less Ghost Dog and more God of War. Like I was using rage to clean residual fear out of my colon, maybe my soul.

## KIRKWOOD, MARCH 20, 2021, FIFTIETH DAY ON SNOW

Day fifty. This was my minimal viable goal for this season, and the hardest physical challenge I'd ever set for myself. Did it feel like victory?

It felt like unfinished business. It also felt like a solo mission to Kirkwood was the only appropriate celebration.

The universe rewarded my efforts. I got to the mountain to discover a foot of fresh snow. The powder meant that I would have the opportunity to right some wrongs—which also felt appropriate.

Three wrongs in particular. First, Main Nostril, the site of my dreaded T-boning. Second, I needed a return trip through the Fingers, as the last time through had been the time Ryan screamed, "Halt!" Finally, the Notch, from the top, which was one of the few remaining lines on my line list.

A low-angle tree run to get a feel for the snow, then a top-to-bottom burner to get a feel for the speed, then I headed to the Wall so my brain could adjust to the steeps. My first lap was sloppy. My second lap was worse. My limbs felt heavy. My brain felt sluggish. Was the weight psychological or physical? There was one way to find out.

Next up, Main Nostril.

I was calm on the lift, collected on the traverse, but as I pulled into the chute the twitchiness arrived. When skiing the scary alone, I have a single rule: Never stop turning. Stopping opens the door for the vertigo, which opens the door for shame, which explains the rule: Never stop turning.

Yet so much new snow had fallen that the straight line was no longer a straight line. I only had to point my skis straight for the final fifteen feet. Then I did a second lap just to make sure.

Next up: the Fingers. Miny was calling my name; I wanted that skiing-through-a-cave feeling. Three turns in, the cave disappeared. The new snow was thigh deep. I skied Miny in a whiteout. I skied Miny by braille. I roared as I shot out the exit.

Finally, the Notch, from the top.

The top of Kirkwood is a giant cliff band. The Notch is a notch in that band, fifty feet wide and a couple hundred feet long. A large cornice rims the entrance. Success means airing sideways off the cornice, making a fast right slash upon landing, then a quick left to set up the exit. Miss either of those turns?

Best not to think about that.

All season, when Ryan and I had been skiing the Notch, we'd actually been cutting in below the cliff as there hadn't been enough snow to come in from the top. Today, I got up there and discovered plenty of snow. Too much, in fact.

The new snow had formed a pair of wind lips that ran horizontally across the chute's entrance, one below the other. Skiing the chute now required a triple stager: jump off the cornice and onto the first wind lip, jump off the first wind lip and onto the second, then leap from the second and hope for the best as that landing was completely blind.

Nope. Not going to win this war.

I vetoed the Notch and headed to the terrain park. After all, it wouldn't be a proper day fifty without a few park laps.

Of course, getting down from the Notch was a bit of an epic. Whatever. It wouldn't be a proper day fifty if I didn't suffer along the way.

Plus, after all that suffering, the terrain park was a relief—which was the point.

Grit is a limited resource. Any adventure into Gnar Country demands persistence and resilience, and this uses up our supply. Plus, even if you try to conserve grit—as I discovered on the Notch—stuff is gonna go wrong. But if grit is the only psychological tool you rely upon in times of trouble, then burnout is the likely result.

Put differently, the older version of me would have been so pissed about getting his ass kicked by the Notch that I would have forced myself to ski that line again or gotten off the hill in frustration. The new version chose to lateralize into a significantly less stressful situation—otherwise known as the terrain park—which decreased threat levels, increased skiing confidence, and retuned the challenge-skills balance.

Flow was the result.

I did ten laps. I nailed all my tricks. I wanted to keep on nailing, but I had started underjumping my hits and the rules are the rules. I was too tired. I forced myself to get off the mountain.

How did I know my Gnar Country experiment was working? Fifty days on the snow and I was still thirsty for more.

# Chapter 7

**HEAVENLY VALLEY, MARCH 21, 2021, FIFTY-FIRST DAY ON SNOW**

The lower mountain was frozen, the upper mountain was crowded—except for the woods. I hid in the trees for twenty laps, playing an old game for the first time this season. After two warm-up laps, I skied three different lines, six times each. Each time I skied a line, I tried to ski it faster, using less energy and getting more air.

In other words, my morning began with a *deliberate practice* session, following Anders Ericsson's "repetition with incremental advancement" advice on expert performance. It became a *deliberate play* session—that is, "repetition with incremental advancement that maximizes opportunities for improvisation"—with the increase in speed being the liberating factor.

Ski equipment is designed to work at speed. If you're trundling along at ten miles an hour, you'll never notice this fact. Yet there's a phase shift above twenty-five miles per hour. Energy input goes way down; equipment output goes way up. At speed, a hard carve requires little more than a one-inch ankle tilt, and, seriously, who knew ankles had that much power.

And this brings us to a paradox. As skiers age, many slow down, believing that slower speeds are safer speeds. The truth is, for older skiers, speed is your friend. Far less work and far greater precision means significantly less wear and tear on the body. But tell that to the amygdala.

Pattern recognition resolves this paradox. After three trips through a line, my pattern-recognition system had hardwired the landmarks. Now it was about turn amplitude—bending the skis to initiate the turn, getting light on my feet as they snap straight to accelerate out of the turn. Finally, I started hunting places to turn that acceleration into airtime. After five laps, deliberate practice became deliberate play and from then on it was the gravity dance—fast geometry at maximum velocity and all was right in my world.

Or, all was right in my world until I leapt off a boulder, overjumped my landing, going fat to flat and jamming my pole into the snow on impact. My still-sore shoulder started singing in pain. The voice in my head said: "Well, shit, man . . ."

### KIRKWOOD, MARCH 24, 2021, FIFTY-SECOND DAY ON SNOW

Ryan and I got on the hill early. The snow looked like shiny ice shingles hung by mad elves. Horrible conditions.

Still, we played our game. The rules had not changed. Chase the rabbit. What's different now was the playing field. The entire mountain had become our slopestyle course, with new possibilities around every corner.

Or so we hoped.

The trees down low were tricky, a few inches of cold snow in the shadows mixed with thick slush in the sunlight. The slush was grabby, which is dangerous in the trees. We aimed for the shadows.

It worked for a while, but then we decided to check out Lightning, where there were no shadows. Five turns through perfect powder lured me into speed, but then thick slush said otherwise. I got ankle snatched, whipped upside down, and thumped, shoulder-first, onto the hill. It didn't hurt much—which was something of a relief.

An eight-inch metal plate holds that shoulder together, the result of an old mountain biking incident. I had been wondering if the T-boning had refractured the shoulder, but the old plate had been holding this new fracture together. An X-ray would solve the mystery, but the pain was minor, my mobility had returned, and all a doctor would say is, stop skiing, or stop skiing and have surgery.

Not a chance. There was still work to be done.

Toward those ends, Ryan started practicing lazy boy 180s off park jumps, laying back on his skis and dragging his hand in the snow to spin backwards while simultaneously launching off the lip. Think *Crouching Tiger, Hidden Dragon*, but on skis.

Of course, the rules are the rules, so I did the same. My version was a lot less elegant. *Crouching Chihuahua, Hidden Smurf*—but you get the picture.

By day's end, I was pretty sure I had the lazy boy 180 wired and decided to try one at speed. I aimed for a large roller, lay back, dragged my hand, and discovered a hidden pile of slush. The slush grabbed my ski and whipped me into the air. I went up, I came down, headfirst and hard. Did I get to add a minor concussion to my ever-lengthening list of nagging injuries?

Indeed, I did.

## NORTHSTAR, MARCH 25, 2021, FIFTY-THIRD DAY ON SNOW

Keoki Flagg is an old friend. He's a celebrated artist and photographer with a gallery in Olympic Valley, and another in San Francisco. He's also the person I've known the longest in Tahoe, as he was the photographer who shot the very first ski story I wrote about the area—some thirty years ago.

Today, Ryan and I met Keoki at Northstar. It was the first time the three of us had managed to ski together since COVID began. It should have been, could have been, would have been—but the mountain was not cooperating.

Bulletproof ice in every direction. It took five hours of search-and-regret before Keoki discovered a single line through the trees that was worth a damn. We skied it twice before he had to split. Then Ryan and I headed back to the terrain park to see if the jumps had softened in the sun.

We took ten laps. For the first five, I missed every trick I tried. It was frustrating, but inevitable.

Flow doesn't work like a light switch. It's not on or off, in the zone or out of the zone. Flow is a four-stage cycle: struggle, release, flow, and recovery. My frustration told me I was in the struggle phase. The key is knowing that frustration is a necessary part of the process. We need the stress response to prime us for flow. Rather than being a signal that I was doing something wrong, my frustration was a sign that I was on the right track.

On lap six, my struggle paid dividends. I managed to link a couple of turns together and found a little rhythm. My body relaxed, flushing the remaining stress hormones from my system as struggle gave way to release.

Now it was time to get creative. I glanced around the hill. I noticed a side hit I hadn't seen before. My brain mapped the angle of the take-off against my catalog of potential tricks and found a match. The angle looked like it would toss my skis to the right, making it a perfect place to try a safety grab.

See the line, ski the line; that's the rule.

The rule paid dividends. I nailed my best safety grab of the season. Now I was in flow.

Thirty second later, Ryan pulled off his first double truck driver—kicking both skis in front of his chest while grabbing them below the tips with the same motion that a truck driver uses to grab the steering wheel. Now he was in flow too.

In celebration, Ryan slid every rail in sight. I slid nothing but a wimpy pyramid-shaped box, but I didn't fall on my ass again, and that was victory enough. We called it a win and got off the hill.

In the parking lot afterward, Ryan mentioned that earlier in the day, Keoki had watched me ski a line and said, "Steven's a totally different skier."

I didn't know if that was a good thing or a bad thing, but until otherwise informed, I glowed all the way home.

### KIRKWOOD, MARCH 27, 2021, FIFTY-FOURTH DAY ON SNOW

The bitchy snow continued. Thick and slow in the trees, frozen and slick on the runs. We tried for twenty laps, but conditions wouldn't cooperate.

One so-so line down Oops, one solid line down Chamoix, one decent nose butter 360 along the way. The only real progress came from a shifty over an up-down box, which I nailed twice in a row.

Then Ryan underjumped a 540, landed sideways, and torqued his already torqued knee. Next, I caught my tails trying to spin another nose butter, smashed into the ground, and landed directly on my injured shoulder. The voice in my head said, "Go home, buddy, it only gets worse from here."

Who was I to argue?

### HEAVENLY VALLEY, MARCH 29, 2021, FIFTY-FIFTH DAY ON SNOW

It got worse. I woke up the next day, tried to sit up, and the world started spinning. This wasn't residual vertigo; this was my recent concussion.

I took a day off to rest, then headed back to the hill. I figured, take it slow, ski the groomers, just stretch the legs. Ten laps total, then go home. What I hadn't counted on was spring break.

It was Jerry World.

Okay, Jerry, let's review: The chairlift is a slow-moving bench. As the bench approaches your ass, sit your ass down. At the top of the hill, stand your ass up.

Also, please remember, the combination of high-speed action sports and crowded on-the-mountain conditions—absolutely, this situation improves dramatically if you apply large quantities of alcohol.

### KIRKWOOD, MARCH 30, 2021, FIFTY-SIXTH DAY ON SNOW

I hate seeing myself on video. Watching myself is torture, watching myself ski is worse. To date, I've been filmed skiing exactly twice, and this was after almost fifty years of skiing. The issue stems from the first time I was filmed skiing, in Chamonix, France, 1994. I was on assignment, following extreme skiing pioneer John Egan down a steep chute with a small cliff at the bottom. Somewhere behind us was a cameraman. I skied the line, hit the cliff, landed backseat, bounced front seat, smashed face-first into my pole, broke my pole, blacked my eye, and cartwheeled to the bottom of the chute.

When I watched the video, never mind the tumble, the worst part was the approach. There was John Egan making really nice turns down the chute. But the guy behind? That guy was a mess. All hips and shoulders. He couldn't keep his body pointed down the fall line, his turns were swishy, his torso wiggled like a worm. It was Jerry World—only, in this video, I was Jerry.

For this reason, when Ryan suggested we pull out his phone and spend the day shooting videos of me skiing, "Kill me now," was my response.

In my mind, with Ryan shooting video, the project itself was at stake. I wanted to be steazy in the sick gnar. If I saw video of my park skiing and it was still Jerry World—the horror, the horror, the horror would be too much to bear.

The voice in my head began offering words of encouragement: "Hey, candy-ass, chickenshit, limp dick. Don't be such a wuss. What are you, moist?"

I needed help; I needed neurochemistry. Look, up in the sky, it's a bird, it's a plane—it's dopamine to the rescue!

As we pulled into the park, I noticed the park crew had installed a new feature: a corrugated pipe stretched horizontally across the top of the first jump. The idea being to smash the ski tips directly into the pipe, letting the force bend the noses inward. When the skis snap to straight, the power of the recoil pops the skier into the air—or what's technically called a "wallie."

The secret to a wallie is to keep your legs locked together. If not, your feet splay wide; and, in the future, if someone wants to invent a testicle-smashing torture device, my guess, it works a lot like this.

I had never done a "wallie" before, but I'd seen videos and it looked fun and isn't curiosity a wonderful drug? I forgot the camera and skated toward the jump. Westside Connection in the earbuds. The world is mine, at least according to Ice Cube.

I smashed into the pipe, popped toward the sky, and it worked—so well, in fact, that the voice in my head decided to offer a midair suggestion: "Tail-tap the pipe."

The results were tail-tap-ish, but my confidence soared. Afterward, I skied seven more laps for the camera and even found a little rhythm. During my 180s, I was pulling my knees up toward my chest and putting more air under my skis. Damn if it didn't feel steazy.

And there was more good news. In the parking lot afterward, we had a little video review. Sure, I still looked like a novice park skier, but I looked like a stylish novice. There was some steaze in my skiing.

Even better, the video revealed four issues in my form, none of which were insurmountable. I had to lower my stance on my approaches, drive farther over my ski tips during butters, suck my knees toward my chest whenever I aired, and add more amplitude into every trick I was throwing.

What a relief. Sure, it still looked like I was learning my way around a terrain park, but I didn't look bad. On a handful of tricks, I looked pretty good. I could work on my issues. The video evidence was strong: a bit of steaze, already on lock.

### KIRKWOOD, APRIL 1, 2021, FIFTY-SEVENTH DAY ON SNOW

It's a goddamn box in a goddamn kiddie park—six feet long and two feet wide and coated in plastic. So what is so goddamn hard about sliding sideways across it? You come off the jump, turn sideways in the air, land sideways on the box, end of story. As long as you land with your skis flat and your weight forward, you slide across the box and what could be so goddamn hard?

Turns out, nothing. Fucking nothing. Less than fucking nothing. You jump, you turn, you land, you slide, you're back on snow.

Then I did it seventeen times in a row to make sure it stayed *less than fucking nothing.*

### KIRKWOOD, APRIL 3, 2021, FIFTY-EIGHTH DAY ON SNOW

When throwing a shifty, you use core muscles and hip flexors. My core muscles were strong; my hip flexors were not. I'd tweaked my right hip flexor earlier in the season. Today, I tweaked it again throwing a shifty on run one. I tried to argue for nine more runs, then left the mountain in defeat after run ten.

Still, there was a little progress. On the drive home, I relearned an old lesson. If you ever need to end a phone conversation quickly—"I'm about to lose you, I'm heading into a canyon"—that shit works every time.

### KIRKWOOD, APRIL 4, 2021, FIFTY-NINTH DAY ON SNOW

I did something I have never done before, not in a half century of skiing—I took a lesson. The video review had revealed weak spots in my game. How to fix them was a different story.

More importantly, I needed to battle-test my positive opinions about my progress. I didn't trust the video review, and Ryan's evaluation remained suspect. However objective he tried to be, Ryan's a nice guy and a good friend. What I needed was a harsh critic willing to call bullshit on my bullshit.

To these ends, I used a classic Jerry maneuver to set up my lesson. When I called up the Kirkwood ski school, I asked for an instructor who can help me nose butter, nollie, stuff like that. "Also," I said, trying for casual, "I need someone who knows the names of all the gnarliest chutes on the mountain—I'm tired of not knowing where I'm going."

"Oh," said the woman at the ski school, "so you need an expert instructor?"

"Yeah," I said, doing my very best Jerry impression. "I'm a real expert. I can ski. I'm just trying to learn how to ski park."

"Real expert" is a term that sets off alarm bells for ski instructors. It's another difference between tourists and locals. Tourists think that "experts" ski the runs marked "experts only" on the trail map. Locals know that real experts also ski the Cirque, the part of Kirkwood that's permanently closed on the trail maps, but is actually open once a year, when the mountain hosts the Freeride World Tour. Plus, a "real expert"? Why would a real expert ever take a lesson?

The ski school would take my statement as a challenge. Of course, on the very off chance that I was telling the truth, which, in actuality, was totally preposterous, but still, on the tiny sliver of a minute whiff of a distant possibility that I was not exaggerating, they would have to find me that expert instructor.

But they would be irked. They'd tell my instructor of my "expert" claim. And while my instructor would teach me stuff I wanted to learn, their real goal would be to teach me how wrong I was about my high opinion of myself, and what it actually means to be a "real expert" at a mountain like Kirkwood, you fucking moron.

The proof that my guise worked? My instructor showed up for our lesson on a pair of Völkl Mantras. I had a pair myself. They're directional skis, not meant for park skiing and never meant for nose butters, nollies, or any of the other tricks I wanted to learn. This meant that while I had told the ski school that I was a real expert with actual park skiing desires, they didn't believe a word.

My instructor's name was Eric Arnold. Turns out, he was an old-school Tahoe punk rocker. This meant, we were roughly the same age, with history and friends in common. Eric's also a real expert.

Back in the 1990s, as a snowboarder, Eric won the Kirkwood leg of the Freeride World Tour on multiple occasions. He was also an expert skier. In fact, he was *the expert*—meaning, the expert that Ryan and I had been watching ski for the past two years.

Every now and again, we'd catch a glimpse of someone in a blue ski instructor's jacket throwing a shifty over a natural double or slicing a precision turn into a burly straight line. We didn't know their name. We just knew they were steazy in the sick gnar. But as soon as I saw Eric ski—I recognized his form.

Then Eric, the real expert, took me, the fake expert, down Oops and Poops for our first run. This was not surprising. Oops is a great test run. It has a little bit of everything, plus it has escape routes. If I wasn't the skier I claimed to be—which is usually how these things go—there were plenty of places to bail out of trouble.

Of course, just taking a lesson pushed me out of the challenge-skills sweet spot. I was too nervous and skied like crap. Of course, I tried to show off by following Eric into a cliff zone. Of course, I hit an ice bump and spun onto my ass.

Whatever. After spending much of the past year falling on my ass in the terrain park, falling in front of Eric was just more of the same.

Next, we skied a line down Cliff Chute and my turns started to improve. Then, we headed to the top of Norm's Nose, and Eric gave me the information I'd been seeking. He pointed out Kirkwood's biggest lines, the mountain's actual expert terrain, showing me what was skiable, telling me when it was skiable, and breaking down the hazards I might encounter along the way.

I stored this data in a mental file marked: "Maybe Next Year, Maybe Never."

Finally came Eric's real test: the exposed steeps of upper Norm's Nose and the skinnier chutes—the Nostrils—below. We blasted down the upper Nose and Eric headed for a straight-line chute I'd never skied before. Yet my return trip to Main Nostril had paid emotional dividends. The fear was gone. Today, I dove into the straight line, skied out the other end, and saw Eric blink and reassess my abilities.

"Okay," he said smiling, "let's go try the park."

On our way over, I felt jittery. It was game time. I was gonna have to do scary stuff. It was a show-and-prove kind of thing. If I wanted Eric to honestly assess my park skiing skills, I would have to push myself to the outer edge of those skills.

I decided to start my line with a ten-foot-long up-down box. It was the longest box I'd ever attempted to slide. And then something strange occurred: I jumped onto the box, slid sideways across it, and came off switch, no problem. Then, I threw a solid 180 off the next kicker for emphasis. Like that ever happens. . . .

The rest of my line was not nearly as elegant. I followed Eric through the jumps. He threw a series of enormous shifties, each with a latched-on mute grab—that is, reaching his right hand across his body to grab his left ski. I attempted three nose butter 360s, none of which had much air under them.

But we did another lap, and another, and another.

Along the way, Eric got my nollie 180s dialed. I'd been trying to learn this trick all season. After he explained the mechanics, I got it immediately.

My real breakthrough came on the jumps. Eric noticed I wasn't getting nearly enough pressure on the tongues of my boots as I was riding up the lip. He repositioned my stance and this turned out to be the very thing I'd been hunting. Suddenly, I was getting a couple extra feet of pop. This led to my first "latched-on" safety grab, where I held my grab from takeoff to landing.

The real victory came on our final line. The last hit on the big line was a twenty-five-foot gap jump. Underjumping it would mean crashing into the backside of the landing, which is why I'd never attempted it before. The secret is speed. If you know how fast to go, you can clear the gap. Eric said he would tow me in—meaning, I tucked in behind him and skied the same line while matching his speed.

Eric took off and I tucked in. Shifty with a tail tap over the first box, 180 off the side hit, then the jump line. Shifty with a latched-on safety over the first two hits, decent nose butter 360 over the third, then we straight-lined for the gap jump. I surprised myself and threw an old-school backscratcher.

And here's the thing: I hit every hit, nailed every trick, and even cleared the final gap jump. I started my season with one big goal—ski a steazy line through a medium-size terrain park. Mission accomplished.

Afterward, Eric looked at me and said, "You are what you say you are."

At first, I didn't know what he was talking about. Then I remembered my Jerry claim to the ski school. "You mean," I asked, "I'm an expert skier trying to learn to ski park?"

"Yeah," he said.

"Yeah."

"That doesn't happen very often."

"I'm still just a beginner terrain park skier."

"No," he said. "Seriously, I've been watching. You're not bad."

Fifty-three years old—so this is what victory feels like.

## HEAVENLY VALLEY, APRIL 5, 2021, SIXTIETH DAY ON SNOW

I did twelve laps while staring at the staggering beauty of Lake Tahoe and composing a poem about the coming of spring:

*Snow melts off the mountain.*
*Soon they will shutter the ski resorts.*
*Fuck, fuck, fuck, fuck, fuck.*

## KIRKWOOD, APRIL 7, 2021, SIXTY-FIRST DAY ON SNOW

Today, the melt-out was in full swing. We saw the damage everywhere. Recently, snow-coated boulevards had been replaced by thin, dirty ribbons. Once-white bluffs were now completely exposed rock.

I felt the gut punch of time. The season was drawing to a close. There would be a pause in my progress and an end to the easy access to flow that progress provided. The easy access had been keeping my nervous system in check. Yet, without the volume control that came with flow, the voice in my head would only get louder.

"I'm not ready to be off my meds," I told Ryan on the chairlift, pointing at one of those now-bare bluffs.

We fought back. Skiing fast, taking chances. Point of fact, we skied Cliff Chute twice, as our warm-up.

The Fingers were next. But bad vibes up top. We decided to do something we'd rarely done all season—think first, act second.

We skied around the chutes to inspect them from below. It was a decision that kept us out of the hospital. The bottom half of both Eeny and Meeny had completely melted out. About the time we would have reached terminal velocity, we would have run out of snow.

The last two chutes, Miny and Moe, appeared to be skiable. Were they actually skiable?

We lapped Miny first. Steep, narrow, technical, and that skiiing-through-a-cave feeling never gets old. Next up, Moe. But something strange happened at the top of Moe. Ryan, who never gets spooked, got spooked.

It was the snowmelt. There were exposed rocks everywhere. The only skiable line looked even skinnier than it had during our last trip. It also looked like some sketched-out snowboarder had side-slipped the entire thing, pushing off all the snow and leaving behind a rough staircase of bulletproof ice.

But—for whatever reason—I saw the line. "Let me lead," I told Ryan. "I definitely owe you."

After one awkward entrance maneuver, three gnarly hop turns, and a thirty-foot straight line exit, that grudge was settled.

Next up: the terrain park.

Of course, I had to show off my new box-sliding skills. Box slide to forward, box slide to switch, and so forth, through the park.

Even better, in the jumps, my incremental progress continued to incremental along. I was getting more pressure on the tongues of my boots and more air off the jumps. On my second lap, I threw a pair of latched-on safety grabs over the first two hits, which, according to Ryan, even impressed the gaggle of teenage girls gathered up top.

First time in thirty years that's happened.

We spent the next three hours lapping the park. Twenty-four laps later, as I got ready to throw a nose butter 360 off the final jump, I ran out of gas. I was too tired to lean over the noses of my skis. But my brain had an unusual suggestion: Don't lean. Just look. Pop into the air, twist your head, and spot your landing.

And this is how I threw my very first 360. I jumped, looked, and landed. It was a full-fledged 360. It was the very trick I'd wanted to learn for nearly forty years, the very trick I'd been scared to do for nearly forty years. I did it all right, and it was totally accidental.

Then I did it again.

### HEAVENLY VALLEY, APRIL 8, 2021, SIXTY-SECOND DAY ON SNOW

With my focus on the terrain park, I sometimes forget there's another activity taking place on the rest of the mountain. From what I can gather, it requires waterproof clothing, and sliding down snow in long swooping arcs.

As the great prophet Too Short says, "Blow the whistle."

### KIRKWOOD, APRIL 10, 2021, SIXTY-THIRD DAY ON SNOW

My plan was to stretch in the parking lot, stay mellow until things softened in the sun, then get after it in the afternoon slush. But Eric Arnold pulled into the lot just as I finished stretching and asked, "Did I have any interest in park laps?"

So much for mellow.

Our posse started out with Eric and myself, but Ryan joined after lunch, and throughout the day, we'd link up with different groups for different runs that all had one thing in common—speed.

It was Kirkwood's last weekend, and nobody wanted to waste a moment. Everybody was in full send mode. Everybody dropped into group flow as a result.

I was shocked by the result.

While I regularly dropped into group flow with Ryan, finding myself in the state with a collection of strangers was a much rarer experience. I'm an introvert with an injury-riddled past. I don't like the sense of competition that emerges in groups. My preference is to ski with

Ryan, taking a collaborative approach to progress and deploying group flow's most potent trigger: "Yes, and."

"Yes, and" means Ryan and I aim for cooperative and not competitive. When we play "yes, and" games like Chase the Rabbit, I never try to beat Ryan's performance. Rather, I want to build on his ideas, making our skiing cumulative (adding to what he does), and not combative (trying to outdo what he did).

But deploying "yes, and" takes practice. It requires trust. It demands checking your ego at the door and aiming for equal participation in everything you do. In other words, it's unlikely to happen when a posse of amped-up locals gets together for the first time at a mountain like Kirkwood.

But that's exactly what happened today.

The highlight was Devil's Corral, an area directly below the Cirque. I'd never been in there before. Steep faces that funnel into roller-coaster chutes between huge rock walls. A year ago, this place would have scared me to death. Today, I skied it with a posse of ex-pros and held my own. This was no small victory.

Even though it had never been a stated goal, even though, for an introvert like me, saying this aloud would have been more crack pipe dream than pipe dream, this was why I trained so hard in the off-season and worked so hard this season. On any mountain, there's always one uber goal: to be able to link up with a posse of locals and drop into group flow and add something to the mix. That's the real job: Be skilled enough to pour your creativity and courage into the group's collective effort so the group, as a whole, becomes better together.

In fact, when most people heard about my Gnar Country adventure, they thought the impossible challenge was me learning how to park ski. The truth? Learning how to park ski was the hard thing I had to do to get to the real impossible challenge—that is, setting down the shame of my past and the self-consciousness of my present long enough to actually be part of a group.

Or, as Eric Arnold likes to say: "It takes everyone."

### KIRKWOOD, APRIL 11, 2021, SIXTY-FOURTH DAY ON SNOW

I was exhausted. I drove to the mountain anyway, needing to see her one last time.

Ryan was skiing with his kids, so we did park laps together, taking his five-year-old son through his first line of big jumps. The kid's a ripper, just like his dad.

I was not a ripper. Too tired to do much but miss my safety grabs, underspin my 180s, and fall off boxes, it took everything I had to produce a single steazy line. I'll spare you the details, but I was not displeased with the results.

Then we ditched the kids and took one last lap down Main Nostril, just to make sure the memory of my T-boning stayed a memory. Pulling into the top of the chute, my exhaustion started messing with my perception. I saw terror in places there was none.

"Not going out like this," said the voice in my head.

I shook off the bad vibes, cleared my head, saw the line, nailed the line, and nearly collapsed. The tank was empty. I'd left everything on that hill, which is the way I like it, which, in my opinion, is the only real way to honor a mountain.

It's always a trade. The mountain allows you to play in places that human beings shouldn't be able to play. In exchange, you must pour all of yourself into those places, constantly confronting the demons that inhabit them, pushing past your ancient fears, your puny ideas about your potential, even your notion of yourself. In the end, if you get into flow along the way, you disappear completely. In the end, if you do it right, all that remains is the mountain.

Thank you, Kirkwood, it has been an honor and a privilege to make your acquaintance.

### NORTHSTAR, APRIL 14, 2021, SIXTY-FIFTH DAY ON SNOW

One final trip to Northstar's terrain park—at least that was the theory.

The reality: Northstar had razed most of its terrain park. The kiddie line was gone, the small line reduced to two jumps. Only the medium and expert lines remained intact. The expert line was out of the question. The medium line was the same line that kicked my ass on my first park session of the season—and I'd been avoiding it ever since.

Today, I surprised myself. I skied into the medium line with considerable trepidation, but nailed a latched-on safety grab over all three of the jumps, then slid three of the longest boxes I'd yet attempted, then sent a shifty to the moon off the final booter.

And this brings us to another Punk Rock Digression:

The DIY spirit in punk rock was about using collective creativity to make meaning and purpose in the face of bad odds and no other prospects. While this same spirit has long been present in skiing and snowboarding, terrain parks were the ultimate "punk rock" innovation—as they allowed everyone to play.

Until terrain parks came along, expert-level riding demanded expert-level terrain. This meant access to pricey resorts, with their pricey lodging, expensive lift tickets, and other financial barriers to entry.

But anyone with a hill could build a terrain park.

Nearly overnight, working-class kids in Middle America could access the same expert terrain parks as wealthy kids from the West Coast. This supercharged the rate of innovation in freestyle and democratized opportunities for success. For example, X Games gold medalist turned X Games commentator Tom Wallisch came from Seven

Springs, Pennsylvania—a resort three hours from my childhood home in Cleveland and a place that never produced pro skiers until terrain parks came along.

And this ends our Punk Rock Digression.

Back at Northstar, we gave the medium line another go. I got a little overzealous on the final kicker, retweaked my hip flexor, and that was the end of the fun.

Ryan, meanwhile, was on a mission. Three weeks ago, he threw his first nose butter 540—in the kiddie park. Today, he was sending them to the moon off expert kickers. Same thing with his zero spins. Three weeks ago, he was testing the trick on tiny rollers, today he was throwing them off serious booters.

In the parking lot afterward, Ryan was still fired up. I felt too dejected to be fired up. This was the third time in two weeks that I'd retweaked my hip flexor. Sure, the first time I visited Northstar's terrain park, I'd cleared zero jumps and slid zero boxes. Today, I cleared every feature on its medium line on my first go. Yet this didn't cure my dejection. I wanted more. More progress. More time to make more progress. More progress for more flow for even more progress.

Apparently, one inch at a time was no longer fast enough for my ego.

# Chapter 8

**SQUAW VALLEY, APRIL 24, 2021, SIXTY-SIXTH DAY ON SNOW**

I took a week off to rest my hip flexor. It was the longest I'd gone without skiing since the season began. It felt like a chunk of my soul had been amputated.

Compounding matters, my brain began running what-if scenarios. What if this wound won't heal? What if I can't ski again this season? What if . . . ?

To quiet the uncertainty, I got in touch with Fred McDaniel at the Human Performance Center in Santa Fe, New Mexico. An occupational therapist, Fred has spent twenty-five years helping injured athletes return to peak form. About a decade back, he restored mobility to my shattered ankle. Ever since, he's one of my first phone calls when injured.

After reviewing my yoga program, Fred realized my quadriceps weren't getting the attention they required. Compounding this, arguably to compensate for the T-boning damage, my glute muscles had deactivated, leaving my hip flexor to do most of the work.

That explained the damage.

Luckily, the solution was straightforward. We added six stretches to my yoga routine, three for my quads and three for my hip flexors, alongside glute reactivation exercises like lunges and squats and hip flexor exercises like banded clamshells. Additionally, while skiing didn't seem to exacerbate the tear, I definitely had to stay out of the terrain park until it healed.

Today, it was time to test my healing progress.

Yet with the resorts on the south shore of Lake Tahoe shut down for the season, the test required a visit to the north shore. It was time to return to my old stomping grounds, Squaw Valley.

I was nervous. Squaw's terrain is ferocious. I had bad memories. I had years of experience skiing this mountain—poorly.

Still, I had worked hard all season and felt prepared.

I was not prepared. The steepness of the mountain, the size of the moguls, the weight of my memories—all of it messed with my head. Plus, it was icy. Plus, it was insanely crowded. Plus, I skied like absolute crap.

The only good thing I can say about day sixty-six—at least my hip flexor held up.

### SQUAW VALLEY, APRIL 26, 2021, SIXTY-SEVENTH DAY ON SNOW

Sneak attack: A late-spring blizzard blew in overnight. The morning brought high winds, wet snow, and whiteout conditions. Most of Squaw's upper mountain was closed. Lower mountain, only Red Dog and KT-22 remained in play.

KT was a near-death experience. Visibility was less than two feet. The tops of the moguls were sheet ice. The bottoms were a soggy mess. This made for high speed acceleration on the ice, high speed deceleration in the slush, sore knees, wounded pride, and the general desire to get the hell off KT and go investigate Red Dog.

Red Dog wasn't much better, but the bad weather kept the tourists away. We took advantage of their absence, doing high-speed laps down empty groomers, then flipping around switch and doing high-speed backwards laps down empty groomers. Then the temperatures dropped again, and the wet snow turned into dry powder, and no one—besides Ryan, myself, and a handful of hardcore locals—noticed.

By early afternoon, enough dry powder had fallen that the ski patrol opened Red Dog Glade, the double black zone directly beneath the Red Dog chairlift. The glade is a majestic pine forest striped by a series of steep, rocky chutes. From below, the chutes look like gargantuan claw marks—now covered in eight inches of untracked powder.

By midafternoon, Ryan and I were joined by another old friend, Tom Day, who is best described as ski royalty. Now in his sixties, Tom moved to Squaw Valley in 1982, as part of the first wave of extreme skiers to call this mountain home. He started out starring in ski films, then moved to the other side of the lens and became Warren Miller's main camera operator for the next three decades. From *Blizzard of Ahhhs* through *Steep and Deep* to *Future Retro*, most every classic ski film you can think of, Tom either skied in it, shot it, or both.

As a skier, Tom is smooth, powerful, and catlike, but at speeds that are hard to explain. Fast is the short version. On a clock, Tom skis about ten miles per hour faster than Ryan and I normally ski, but it's where his speed comes from that makes the biggest difference.

Tom sees the mountain unlike any skier I know. On our first run together, we skied the same line that Ryan and I had already skied three times. Tom skied almost the exact same line, but set his turns about two feet away from where I'd set mine, in places I would have never considered trying to turn.

Yet, every time, the results were smoother and faster. A lot faster. Ten turns in, everything just blurred. "Full send," as the kids like to say.

Full send sent us straight into group flow. It appears that my new-found skiing confidence had lowered the amount of stress hormones I produced in the company of skiers not named Ryan. Normally, these hormones smack me out of the challenge-skills sweet spot, rob me of my ability to automatically execute motor patterns; thus begins the cycle of shame.

But not today.

Today, my group flow experience at Kirkwood was paying emotional dividends. Today, Tom, Ryan, and I chased each other around Red Dog Glade until the lifts closed, catching the last chair for a final run through that ancient forest, knowing that an uncrowded powder day at Squaw Valley was a kind of miracle, saying thank you the only way we know, by playing fast geometry at maximum velocity.

Ghost Dog mode, the gravity dance, full send.

### SQUAW VALLEY, APRIL 27, 2021, SIXTY-EIGHTH DAY ON SNOW

Since arriving in Squaw Valley, Ryan and I have been crashing at Keoki Flagg's house. Keoki, meanwhile, has been out of town, on an annual backcountry mission to hike and ski the big peaks in the southern Sierras.

Today, he drove back into town so we could spend the day together, knowing that Squaw was about to shut down for the season, knowing that this would be our last chance for a last dance.

We were excited. If yesterday's dry powder conditions held up, and the whole mountain was open, this could be quite the day.

The whole mountain was open, but conditions hadn't held up. We discovered bulletproof ice mixed with Sierra cement. Tough slogging and I was already exhausted. Of course, Keoki wanted to ski the gnar. Of course, the vertigo set in quickly.

I tried to fight back. A couple of awkward lines down Granite Chief, a couple of uncomfortable lines beneath the Attic, then a KT-22 beatdown—taking frozen ice moguls to the spleen in the steep trees between West Face and Chute 75.

When it was over, I was too tired to do much more than whimper.

Still, it was fun to remember all the terror that waits around every corner at Squaw Valley—in a G-dash-D help me kind of way.

## ALPINE MEADOWS, MAY 1, 2021, SIXTY-NINTH DAY ON SNOW

Near-perfect spring conditions. Temperatures in the high sixties. We wore sweatshirts. We should have worn body armor.

## ALPINE MEADOWS, MAY 2, 2021, SEVENTIETH DAY ON SNOW

It was the last day of the season at Alpine Meadows. The tourists were gone. Only the locals remained. Everybody knew somebody.

We started out skiing fast and kept going. The pack grew. Before noon, I was skiing with five old friends, three new friends, and a random assortment of strangers. So many bodies moving at such high speeds. . . .

By early afternoon, temperatures were in the seventies. The heat increased the snowmelt and the mountain came alive. We skied steep, technical lines through exposed cliffs. Each turn unleashed a river of slush and everybody had to river dance to survive.

Once again, I found myself dropping into group flow. Oddly, over the past month, this shared state had become so familiar that I didn't even notice. Then, Ryan, myself, and another old friend, Gordon Fields, spun simultaneous sliding 360s across the side wall of a steep gulley. Then, I noticed.

Ryan noticed as well. He saw Gordon spin and gave me a tiny nod. I knew that nod. I knew what he was saying.

Gordon was in his late fifties. Over the course of the season, Ryan and I had skied with a bunch of my friends who were this age or older. None of them had a park skiing background. Most of them had never considered park skiing. Most started out thinking my quest to learn to park ski was insane. But after spending a day watching Ryan and me twirl around the mountain, each and every one of them tried a sliding spin 360 of their own.

And then another and another. . . .

It appeared that our Gnar Country approach was addictive. Once people realized there was an accessible entrance point to park skiing—the sliding spin 360—their fear was replaced by curiosity, and dopamine did the rest. The neurochemical amped up pattern recognition, fast-twitch muscle response, and a willingness to take risks. The result: Someone who had never done a sliding spin 360 nailed one on their first try.

We kept twirling around the mountain until the lifts closed, then hung around the parking lot afterward, telling stories, laughing hard, and, of course, making plans for more skiing. Then I walked back to my truck and gazed up at the mountain. It would be six months before I would ski here again. It made my heart ache.

I said goodbye to the snow. I said goodbye to the trees. I said goodbye to the mountain. I thanked the snow and the trees and the mountain for keeping me safe and teaching me more about myself than I can learn on my own. I thanked the mountain for obliterating my puny ideas about my limits. I thanked the snow for facilitating this obliteration. I thanked the trees for keeping me company along the way.

Then I got in my car, cranked up Five Finger Death Punch, and drove on home.

## MAMMOTH MOUNTAIN, MAY 4, 2021, SEVENTY-FIRST DAY ON SNOW

Ryan and I headed south, to Mammoth Mountain. Three weeks had passed since either of us skied through a terrain park. We both felt rusty—which was not what you want to feel when heading to Mammoth.

Mammoth sits at the epicenter of the park skiing world. True to its name, the mountain sports five different terrain parks and over one hundred and fifty different features, most of them ferocious. Many of the SLVSH videos I'd been watching were filmed at Mammoth. And this was the real reason we'd come—it was time to test our skills.

But first, we needed to shake off the rust. We decided to go head-on at that problem, if, by head on, you mean skiing directly to the kiddie park: two small jumps and a friendly beginner box.

I have my rules. Ryan has a few of his own. One of Ryan's rules involves not doing what everybody does—skiing into a terrain park and stopping at the top to "theoretically" scope out a line.

Ryan's Rule: Never stop at the top.

If I was keeping office hours, I'd explain this rule via new theories about extended cognition and the action-perception cycle. Out here on the mountain, all I can say is: Ryan's rule works.

Here's why: If I ski into a terrain park and stop, my access to the full menu of affordances declines dramatically. But if I blast in and don't pause, heading toward the first feature that piques my curiosity, the one-two punch of rhythm preserved and dopamine released gives me a better shot at flow.

Today, I followed the rules. I sliced into the park and aimed for the first jump, and the voice in my head decided to make things interesting.

"Safety grab."

The gauntlet was thrown. The voice was suggesting I make my first attempt at a safety grab since injuring myself during a safety grab. All I can say about that attempt—at least I didn't injure myself again.

The first lap was ugly. The second as well. During those laps, I remembered something that I had forgotten during my three-week break: park skiing—not for the faint of heart.

Then, Ryan and I left the kiddie park and decided to visit Mammoth's main park, which is creatively named Main Park. Forget the faint of heart. Main Park was downright terrifying.

Of course it was. Mammoth prides itself on building mammoth

parks, and those parks always attract the best athletes. Today was no different. What does it look like when a line of seven professional snowboarders and twelve professional skiers hits a four-pack of fifty-foot booters?

Like nothing you've ever seen.

Then we realized what we were seeing. They were filming a game of SLVSH in the Main Park. We couldn't identify all the contestants from the chair, but the signature style of SLVSH cofounder and Olympic gold medalist Joss Christensen was clearly visible. He was the one tossing 900s around like Frisbees.

We put the expert fare of Main Park on pause and instead took a lap through Forest Trail, the medium park. It was more crowded than Main Park, which helps explain why I missed my safety grab on the first jump and underrotated my 180 on the second and no time to dwell, as there were a pair of snowboarders hot on my tail.

Then the line split and my choices were turbo rail or metal ball. The turbo was above my pay grade, but the ball was intriguingly weird: four feet in diameter, perched atop a sloped, six-foot pedestal of snow, and painted hot pink. Theoretically, riders use the pedestal as a ramp and the ball as an object to bounce off. No, thank you. Instead, I shiftied off the side of the jump and tried to nose-bonk the ball.

No bonk. But it was fun to try.

In the jump section, Ryan sent a 360 off a twenty-foot booter, while I launched over the knuckle and tried to tail-drag my skis on the downslope. It was more of a tail tap than a tail drag, but it was the first time I'd tried to tail-tap a big knuckle and hello, Ghost Dog.

A flow state is a terrible thing to waste—that's another rule. So what else can I try? So what else can I learn? I was hungry. I was thirsty. Also, I was in flow, so I was pretty sure I was bulletproof.

Turns out, I wasn't bulletproof—or, more specifically: snowboarder proof.

The park finished with a mini-halfpipe. The voice made an interesting suggestion: tail-drag the lip of the pipe, then sliding spin 360 down the wall—a supersteazy combo if done correctly.

Unfortunately, I lost too much speed on my tail drag and got caught by the snowboarders as I set my spin. One blasted below me just as I started to twirl, another jumped beside me as I came off the lip. So no supersteazy combo, after all.

And the crash—that one hurt.

So, you know, welcome to Mammoth.

### MAMMOTH MOUNTAIN, MAY 5, 2021, SEVENTY-SECOND DAY ON SNOW

The intimidation continued. The park was crowded. Skiers and snowboarders taking off every ten seconds. I had a tiny window to drop and don't stop, as I was being chased by some of the best athletes in the world.

Point of fact, on my first three laps, I was either being tailed by or trying to tail Eileen Gu, a eighteen-year-old phenom and the first female rookie to medal in all three X Game freeskiing disciplines in the 2021 games, before winning two golds and a silver at the 2022 Beijing Olympics. Thus, while I was aiming for a latched-on blunt grab over a small tabletop, one of the best skiers alive was trying a "front-swap, back-swap, 400 out," off the turbo to my right.

No surprise, I never did nail that blunt grab.

After five laps in the park, I realized it wasn't going to work. I had post-traumatic snowboarder disorder. I was seeing ghosts.

Ryan was heading in the opposite direction. He sent a cork 360 off a twenty-five-foot booter, than tried a cork 540 off the next. He almost nailed the 540, but underrotated a tiny bit and wanted another go. This time, he blasted off the lip, overrotated his spin, and had to open up like a starfish—that is, spread his arms and legs wide to dump speed and halt rotation—to save himself from twirling into oblivion.

While Ryan landed without incident, watching his starfish maneuver scared the pants right off me.

And, speaking of pants-less activities, this being springtime at Mammoth, we also saw five grown men skiing in nothing but their underwear, including one very well-tanned individual in a shimmery bronze thong.

Still, despite the thong and the throngs, by day's end, I found a little rhythm. I managed a string of the biggest 180s of my short 180ing career, pulled off a tail press on the lip of the mini-pipe, and finished with a sliding spin 720—which was the very first double I'd ever thrown and perhaps the lamest double anyone can throw.

But progress is progress and, lame or otherwise, I'd take it.

### MAMMOTH MOUNTAIN, MAY 8, 2021, SEVENTY-THIRD DAY ON SNOW

I zipped home for a few days, then came back to Mammoth for a week-long solo mission. A week in their terrain park to tick off the rest of the tricks on my list.

I was chasing a dying dream. Soaring temperatures had accelerated the snowmelt. On my lift ride to the top, I saw that Mammoth had disassembled most of their terrain park, harvesting the snow from one area to keep a different area alive for a few more days. In my absence, they'd razed the mini-pipe and every jump in the medium line. The large metal ball was gone. The wall ride was gone. There were no more side hits or beginner boxes or anything vaguely friendly. All that was left was a long line of mean rails.

I thought about those rails, but decided to warm up in the kiddie park. It was in worse shape than the main park. The boxes were gone. The two jumps were sagging. I didn't care. I applied Ryan's rule and roared toward the first jump, dead set on setting the tone for my entire trip with a huge 180.

What I discovered midair: The jump's backside had melted out.

While I did throw a 180, I landed in a hole, tweaked my knee, and lost my mojo.

Those mean rails would have to wait.

## MAMMOTH MOUNTAIN, MAY 9, 2021, SEVENTY-FOURTH DAY ON SNOW

I woke up feeling old and tired. My tweaked knee hurt. My T-boned shoulder ached. But the snow wasn't going to last much longer.

I decided to go exploring, hoping that the novelty might drop me into flow, and flow might fix my mood. I investigated every run off Face Lift Express, then made my way to a series of berms carved into the far edge of West Bowl. Ryan had pointed them out earlier in the week. The berms looked enticing, but they ran along the edge of a big cliff and fear of vertigo had kept me away.

Yet, on closer inspection, the berms were huge and my chances of dying seemed minuscule. High perceived risk, little actual risk, and welcome to the challenge-skills sweet spot. Plus, those berms produced serious G-forces, and the novelty of that sensation acted as another flow trigger. The combination dropped me right into the zone.

Once there, I decided to get some work done. Ten rollers were carved into the backside of the Face Lift Express. On the first roller, I forced myself to do the thing my body had stopped wanting to do: lean far over my tips to initiate a nose butter 360.

The results were better than expected. I upped the ante, trying to turn the rest of the rollers into a mini-slopestyle course: two decent 180s, two solid nose butters, and an old-school spread eagle off the final roller in celebration. I felt good. I felt alive. But I didn't want to push my luck, so I got off the mountain.

Still, I started my day feeling old and tired and ended it feeling young and refreshed, and this is another thing I've learned this season. A bunch of sensations that I once classified as "feeling my age"—that is, muscle aches, bone creaks, and general exhaustion—seem to be about an increase in performance anxiety rather than a decrease in physical abilities. Once I face my fears and right my mind, my body always makes a comeback.

### MAMMOTH MOUNTAIN, MAY 10, 2021, SEVENTY-FIFTH DAY ON SNOW

I spent the morning trying to add style to my backwards skiing game. Then the temperatures soared and the snow got grabby. I got off the hill in a hurry. Sometimes coming home in one piece is progress enough.

### MAMMOTH MOUNTAIN, MAY 11, 2021, SEVENTY-SIXTH DAY ON SNOW

Will Kleidon is a good friend and snowboarder with a skateboarding background. He drove into town so we spent the day riding together. What caught my attention most was our similarity in line selection.

Most people prefer the center of the run, where there's more space and fewer obstacles. But in the 1990s, riders like Will—that is, snowboarders with skateboarding backgrounds—discovered the edges of the trail, where packed snow berms and raised snow walls offer a myriad of possibilities for improvisation, and thus, faster entry into flow.

Will and I spent the morning investigating those edges. It got a little hectic. Everywhere I went, Will was trying to go. Everyplace he decided to go, I was already there. This is not how things normally work.

Skiers face forward and stand on two independent platforms. Snowboarders stand sideways on a single platform. Typically, these different body positions result in a significant variation in line selection.

Thus, I also spent my morning wondering: *Am I now thinking like a snowboarder?*

One thing for certain, I was thinking kinesthetically for the first time in my life. When I survey the mountain, I now see an entirely different menu of affordances—that is, opportunities for action—than when the season started. Once I get off the hill, those affordances stick with me. These days, I fall asleep watching my brain solve complicated movement problems in the same way as, when I'm writing a book, I fall asleep watching my brain solve complicated wordsmithing problems.

At the start of this project, one of my main goals was to increase my opportunities for creativity while skiing. If I saw an interesting feature on the mountain, I wanted to have the skills to do something interesting with that feature. What I didn't realize was that learning those skills would change my perception of reality. But my foray into park skiing had rewired my brain and altered my affordances.

I'm not just a different skier; I now see a different mountain.

### MAMMOTH MOUNTAIN, MAY 12, 2021, SEVENTY-SEVENTH DAY ON SNOW

Sometimes, the only thing to do is the stuff that scares you most. In exposure therapy, this is known as "flooding." At Mammoth, what scared me most was exposure.

The top of Mammoth is a gargantuan wall of rock. Actually, it's two walls separated by a deep valley. To get up there, either ride the gondola or Chair 23, both of which soar over that valley from seriously unreasonable heights—and that was my issue.

While I'd skied here about two decades ago, I'd forgotten about the exposure. When I came back to ski the terrain park—which sits lower mountain—I didn't pay the top much mind. But now, with the increasing snowmelt, all that's left of the park are huge rails and monster booters.

Thus, I decided to see if I could negotiate with my vertigo.

There was no way to not feel terror on the gondola ride to the top, but I had to keep the fear in check. If I freaked out going up, getting down with the world spinning was going to be a much bigger problem. Worse, every line up top had a cornice.

My problem wasn't skiing off the cornice. My problem was that before the skiing could take place, I had to venture to the edge of the cornice, peer into the void, and figure out where to ski. With vertigo, peering into the void is always a bad idea.

This was where the rule—see the line, ski the line—comes into play. Motion was the only way to deal with cornice-induced vertigo. I had to ski to the edge of the cornice, immediately spot my line, and start skiing before the vertigo could hit.

This is also where repetition suppression returns to our story. By simply riding the gondola and living through the experience, repetition suppression would reduce my fear response the next time around. Eventually, it would extinguish it completely.

It took five unpleasant laps.

First lap, I rode the gondola, shut my eyes, and did deep breathing exercises to keep my nervous system in check. Whenever I felt calm enough, I opened my eyes for a second. I wanted my brain to experience the exposure, but not enough to overwhelm my system.

I skied Cornice Bowl to the bottom—a mellow, single black diamond. And while the bowl does have a cornice, it's minuscule. It only took a second to glance over the edge and spot my line. Three turns in, my brain recognized the familiar motion, realized I was safe, and downgraded the threat level.

Second lap, I kept my eyes open on the gondola ride over the lower mountain, then closed them for the final ascent. I skied Cornice Bowl again, this time jumping off the cornice. Not much more than a six-foot drop, but that feeling of flight never gets old.

Third lap, I battled anxiety with curiosity. Neurochemically, anxiety and curiosity have the same basic ingredient—norepinephrine—yet sit on opposite ends of the emotional spectrum. A little bit of norepinephrine makes us curious; a lot makes us anxious. But it's easy to transform one emotion into the other, which allows us to use curiosity to block anxiety.

This time, when I surveyed the mountain from the gondola, instead of noticing all the stuff that scared me, I focused on route finding. What line looked interesting? What were its major landmarks? What obstacles would I have to avoid on the way down?

Dave's Run looked interesting. I hadn't ventured over there yet.

But the venturing took longer than expected and, now that I was here, where was the entrance to Dave's Run?

All I saw was a rotting cornice and a series of CLOSED signs. Only a small section of the run was still open, but it had a blind rollover about a hundred feet into the line. Were there cliffs below that rollover?

I was alone. I was twitchy. I made a deal with myself: I would ski the first hundred feet slowly, then peer over the edge of the rollover. If I had gotten lost and ended up above a death-fall cliff, I would hike back out. If I needed to hike all the way back to the gondola and ski the other way down the mountain, then I would do that as well.

See, voice in my head, no need to panic. We've got options.

Then I skied down a hundred feet and looked over the ledge and laughed. The line was mellow. There were no death-fall cliffs and no need to panic. The turns were flowy and fun. So fun, in fact, I decided to ski it another time.

Lap five, on the gondola ride up, I gazed at the scenery and loved the view and felt not a twinge of fear. Repetition suppression on the job.

Then, on the far more vertiginous Chair 23, I repeated the process. This time it only took three laps. Also, turned out that the Drop Out chutes—which sit directly beneath Chair 23—were some of the best turns on the mountain.

Final score: Steven 1; Vertigo 0. Mission accomplished.

### MAMMOTH MOUNTAIN, MAY 13, 2021, SEVENTY-EIGHTH DAY ON SNOW

I was emotionally beat-up from yesterday's fear-extinction laps, but needed to make sure that the dead stayed dead. I got back on the mountain and took one lap down everything I had skied the day before.

Then I was sure.

### MAMMOTH MOUNTAIN, MAY 14, 2021, SEVENTY-NINTH DAY ON SNOW

Real life attempted to assault ski dreams. It took about fifteen laps—but I beat real life into submission.

### MAMMOTH MOUNTAIN, MAY 15, 2021, EIGHTIETH DAY ON SNOW

I got on the chairlift before 8:00 a.m., thinking it had been a warm night and maybe the snow was already soft. I was wrong. The hill was glistening with ice.

Then the lift passed over the terrain park and I noticed a new rail: a "turbo" in the parlance, or a rail shaped like a long, fat sausage. And this turbo looked friendly—and wasn't that interesting?

Rails hadn't looked friendly in a while. Nearly every one I've ever tried to slide this season slid me onto my ass. But I had come to Mammoth to work on my park game, and this turbo looked friendly, and you didn't have to tell me twice.

Thus, I spent the rest of the chairlift running visualizations. What it would feel like to approach the jump, pop off the lip, and land on the turbo? I visualized landing with the weight on my front foot. I visualized sliding to the end of the rail. I ran the visualization three times and thought: *You got this.*

But I didn't have this.

I skied toward the turbo, checked my speed exactly as I had visualized, prepared to pop onto the rail exactly as I had visualized, then chicken-shitted in my ski pants and slid right past the rail.

Apparently, I had a courage issue. But courage is essentially dopamine, and dopamine was available. Hello, Gravy Chute.

Gravy Chute is a single black diamond under normal conditions, but most of the snow had melted out of its entrance, leaving behind only a solitary ribbon winding between big rocks. It was exposed. It was gnarly. It was perfect for my needs.

I skied Gravy Chute, got some courage, got back on the lift, and headed to the terrain park. This time, I skied up to the jump, popped off the lip, landed sideways on the turbo, and—holy crap—I was actually sliding a rail.

And then I wasn't.

Two feet from the end of the turbo, my front foot slipped out, my legs split wide, and I ass-crashed onto the rail. The impact knocked me face-forward into cold steel. POC ski helmets—they work as advertised.

My body was sore from the crash. Not as sore as my ego.

I skied through the rest of the park looking for redemption. The last feature was another turbo rail, slightly larger than the first, but seriously, that first turbo, I had almost slid the entire rail. If I remembered to keep my thighs pressed inward, I would have made it off the other side, end of story.

This time, I skied up to the turbo, hit the jump, and landed on the rail. But with all of my attention focused on keeping my legs pressed together, I'd forgotten to land with my weight on my front foot. Instead, I landed back foot, flipped upside down, and once again ass-crashed into the rail.

Nope, nothing friendly about this rail either.

Final score: Mean Rails 2; Steven 0.

### MAMMOTH MOUNTAIN, MAY 16, 2021, EIGHTY-FIRST DAY ON SNOW

Ryan drove down to Mammoth. I was still bruised from yesterday's rail beating and had little interest in a return trip to the terrain park. Instead, we went exploring.

Three laps later, we spotted a single line of slush bumps running along the far edge of a deserted canyon. The moguls were rotting. The snow was dirty. The cliffs were foreboding. If someone wanted to set a ski scene in a horror movie, this would be the perfect spot.

We took a tentative first lap. The snow was better than expected. Sure, we ran out of snow at the bottom of the canyon and had to hike over lava rock to get back to the main run, but the turns were flowy and why not try a second lap.

By the third lap, we were skiing full-throttle, heart rate heading for the upper dimension, aerobic capacity stretched to the max—which was actually the point. As athletes age, many shy away from high-intensity training and instead prefer long, slow slogs. Those slow slogs appear less physically damaging, yet nothing could be further from the truth.

$VO_2$ max is a measure of our aerobic capacity. For most of the last century, scientists believed this capacity declined 10 percent per decade after the age of twenty-five, and there was nothing we could do to slow the slide. But not so fast. The truth is that $VO_2$ max is another use it or lose it skill. Octogenarian endurance athletes, for example, have been shown to have the $VO_2$ max of healthy thirty-five year-olds. The secret to preserving this capability? Train with hard bursts of intense physical activity on a regular basis.

To those ends, Ryan and I took eight laps down those bumps, each one faster than the last. On lap nine, my knees started to ache. With only five days left to reach my stretch goal of eighty-six days on the snow, I didn't want to push my luck. I called it a win and got off the hill.

During the drive home, the ache became a throb and the voice in my head began to catastrophize. What if this pain is the beginning of the end? What good is $VO_2$ max if you can't use your knees? And if you can't use your knees, then you can't ski, and if you can't ski, then you're a miserable son of a bitch. What if you can't ski and are stuck being a miserable son of a bitch for the rest of your miserable life?

Some days, there's just nowhere to hide.

### MAMMOTH MOUNTAIN, MAY 22, 2021, EIGHTY-SECOND DAY ON SNOW

I took a week off to rest.

During that break, I drove to the Moment Skis factory in Reno for a conversation about another pair of new skis, my Wildcat 101s, which worked well in the park but kept trying to kill me on the hill.

Turns out, my skis had been mounted too far forward. This only allowed me to access half of their edge and none of their sweet spot.

My boots, the Moment gurus claimed, were the bigger issue. The Technicas were designed for big mountain riding, not park skiing. They had zero shock absorption, which helped explain my knee pain.

I switched to a pair of Dalbello freestyle boots. The difference was flex pattern and stance. Instead of the forward lean needed to get my old boots to behave, the new ones allowed a more freeride-oriented, neutral body position. They also had built-in shock absorption, so I wouldn't get knocked out of that neutral position with every bump. Essentially, what all this meant: I'd spent the bulk of my season learning to park ski on the wrong equipment.

Today, with the right equipment, I drove back to Mammoth for a solo mission. The new boots, the readjusted skis, the healed body—after my fourteen lap product test, I was excited by the possibilities.

The real victory was that my knees didn't hurt. I would live to ski another day. Then I got off the hill before my knees changed their mind.

### MAMMOTH MOUNTAIN, MAY 24, 2021, EIGHTY-THIRD DAY ON SNOW

The snow was vanishing. The terrain park was gone. Ryan and I spent the morning trying to get creative with the remaining dregs. Along the way, I started working on blunt grabs—that is, kicking my skis backwards so the tips point at the ground and tails aim at the sky, then trying to grab one of those tails.

In study after study, openness to experience is one of the key personality traits that determines the quality of the second half of our life, I was definitely open to the experience of a blunt grab.

It didn't help. I couldn't seem to latch one.

Eventually, I gave up and went back to practicing safety grabs. A safety grab is like doing a four-foot box jump and a one-handed row with a twenty-five-pound dumbbell simultaneously. Thus, some twenty-two laps later, I collapsed on the side of the mountain and didn't move for quite some time—but I was definitely open to the experience of moving.

### MAMMOTH MOUNTAIN, MAY 25, 2021, EIGHTY-FOURTH DAY ON SNOW

Today was our third attempt to shoot video at Mammoth. Today was also my fifty-fourth birthday. Neither went as planned. Pointing a camera in my direction remains the single easiest way to knock me out of flow. Birthdays always remind me of dying.

I decided to set these issues aside and put all of my energy into the biggest 180 of my entire season. It was huge. It was steazy. Unfortunately, the video camera malfunctioned, so there's absolutely no proof that "it" happened.

And my knees started to hurt again as a result. Also, did I mention, birthdays always remind me of dying.

### MAMMOTH MOUNTAIN, MAY 26, 2021, EIGHTY-FIFTH DAY ON SNOW

It was my last day at Mammoth and my eighty-fifth day on snow. I was one day shy of my stretch goal of eighty-six days. Excluding a possible return to Mount Hood over the summer, this was the end of my season. The lifts opened at 7:30 a.m. I was dressed, stretched, fed, watered, and heading uphill by 7:31.

My knees still hurt, but I skied twenty-seven laps anyway. On a handful of those laps, I skied switch with the best form I've had all season. It felt like a bit of closure. But little else felt closed.

My park game still needed attention. Sure, I'd learned eighteen of the twenty tricks on my trick list, but there was so much left to learn. And my line list felt like a work in progress. While I'd skied all the lines on that list, with the exception of Pencil Chute, none of them felt like big lines anymore. The big lines were the ones I hadn't skied yet. At Kirkwood alone, there were a half-dozen possibilities: Carnell's, Pipeline, Boulevard, the Handrail, Once Is Enough—just writing these names makes me tremble with fear.

Some of my issue was the addictive nature of progress and its sneaky habit of moving the goalposts to make sure the flow tap never ran dry. Actually, all of this was the addictive nature of progress and its sneaky, goalpost-moving ways—but knowing the facts didn't quiet the emotions.

The worst part was the uncertainty. My knees were still an issue. Would the pain go away once the skiing stopped? Was this exhaustion? Was this worse? Once again, the voice in my head argued for worse.

Every second of the drive home, the voice argued for worse.

# Chapter 9

### NORTHERN NEVADA, JUNE 1, 2021

Today was the first Saturday in seven months that I did not go skiing. Today was the first Saturday since Thanksgiving that I did not wake at 3:00 a.m., write for three hours, hike with Kiko, do yoga, eat breakfast, and drive to the mountain. Today, I did not get first chair. Today, I did not play fast geometry. Today, I did not get out of bed.

### NORTHERN NEVADA, JUNE 3, 2021

Weather report: heavy clouds and dark moods.

A week has passed since I last went skiing. Without the regular access to flow the mountain provided, the voice in my head had a *what if* field day. What if I needed knee surgery? What if surgery leads to further degeneration? What if this season was the end of my park skiing career?

To compensate, I devoted myself to recovery—yoga, saunas, long walks, lots of supplements. I needed the swelling in my knees to go down, so I could evaluate the extent of the damage. I needed the soreness in my body to dissipate, so I could head to the gym and evaluate my level of fitness.

Today, while the knee swelling continued, the soreness had abated—so I went to the gym for a little experiment. I wanted to know if my shoulders, back, and chest had lost or gained muscle during the season.

Over the past seven months, in an effort to preserve all my energy for the skiing, I stopped lifting weights. The extra energy made a noticeable difference on the hill, but it was a double-edged sword.

Strength declines with age. As long as muscles stay active, the remaining tissue can often compensate, but it's another use it or lose it situation. If I lost too much upper-body strength by not lifting this season, my fitness approach wasn't sustainable over the long haul. It would mean that next season required a compromise of the less skiing, more lifting variety.

But the results surprised me. Despite my early season rotator cuff tear, I could lift five to ten pounds more on every one of my major shoulder exercises. My back and arms were strong as well. My pull-up count was where it had been before the season, and my dip count was actually a little higher.

My chest was the only major muscle group to have lost strength, but that had a different explanation. The bulk of the T-boning damage had been to my upper chest and lower shoulder, and fear of reinjury made me tentative on the bench press. If experience was any guide, as my fear of injury abated, the missing strength would return.

This meant, for at least half my body, my Gnar Country approach had worked. Of course, the upper half of my body was not the half I was worried about, but it was more proof that tackling hard physical challenges later in life wasn't as impossible as many believed. It was also proof that I could get after it again next season. This was good news.

But it did nothing to change the weather report. I had skied nonstop for nearly seven months. I had become addicted to the progress and the regular access to flow that my progress provided. Now that I had stopped skiing, there was little flow, no progress, and too many questions of the "what if?" variety.

Thus, until further notice, the forecast: heavy clouds and dark moods.

### PINE NUT MOUNTAINS, JUNE 6, 2021

I decided to reintroduce weight vest hikes. I stayed mellow: a forty-minute hike with a twelve-pound vest. My knees felt good. My spirits lifted. Maybe I don't need surgery. Maybe my park skiing career was not coming to a screeching halt. Maybe I could change the weather.

But that's a lot of maybe.

### PINE NUT MOUNTAINS, JUNE 7, 2021

Today, I upped the ante, donning a twenty-pound vest and doing an hour-long hike. I climbed the tallest peak around and my legs held up just fine. But coming down, I felt a twinge in my right knee. Then a tweak in the left. Then a slowly spreading ache in both.

By the time I was back home, I was in pain. The only good news was the swelling had decreased enough during my time off to use this new pain to localize the problem. Before, there had been overall joint discomfort. Now I knew the source of the hurt, which meant I could ask for advice.

I called Fred McDaniel again. We hooked up a video chat and talked pain and sensation. Then he watched me walk, squat, and such. Fred had a tentative diagnosis. It matched my tentative diagnosis. Sure, an ultrasound or MRI would uncover the truth, but I liked to hone my self-diagnostic abilities. Plus, if the truth was of the *you need surgery* variety, I wasn't yet ready to handle the truth.

Still, we both felt sure of our conclusions. I had an inflamed patellar tendon in my right knee, or what's known as "jumper's knee."

Jumper's knee is the product of, well, too much jumping. More specifically, my knee pain resulted from improper training in the preseason and overuse during the season. Preseason, I hadn't done enough sport-specific dynamic movements like box jumps, believing my weight vest hikes would cover this base. In season, I jumped too much and recovered too little, the result of my endurance goal of eighty-six days conflicting with my dynamic goal of learning to park ski.

Next came the T-boning. The damage to my back and shoulder appeared to be overloading my hip flexors. To compensate, my knees took on an extra workload and then the real damage started to accumulate. Finally, I landed backwards in that hole at Mammoth. This didn't help my patellar tendon one bit. Worse, it stretched the medial collateral ligaments in both of my knees.

Patellar tendinitis is painful, but it doesn't weaken the joint. If you can get the inflammation down and strengthen the surrounding muscles, it's manageable long term. The medial collateral ligaments were the bigger concern. I had overstretched these ligaments and there was now fluid in those tears, which was causing both pain and instability. If I could reduce the inflammation, the fluid would drain away and, at least theoretically, I could stabilize the joints through exercise.

But none of this would happen quickly.

Fred gave me an assortment of low-impact exercises to start rebuilding the muscles surrounding my knees. We talked about ways to treat the pain and inflammation. Fred described cutting-edge research and new theories in neurophysiology. I felt myself slipping down a familiar rabbit hole—the one coated in snake oil and miracle cures.

I hung up the phone knowing what that call meant. I had already decided that the best way to progress my skiing in the off-season was time in a trampoline park. A trampoline would let me practice my grabs, flips, and spins. But trampolining was hard on the knees. As was skateboarding. And mountain biking.

That phone call meant that all the high flow activities that I had planned to use as skiing substitutes were still off-limits. It meant, at least for a while, I had no protection against the voice in my head.

"Good luck, buddy," said the voice. "You're gonna need it."

### NORTHERN NEVADA, JUNE 15, 2021

I took Kiko for a hike. My knees still ached. My mood still ached. And without skiing to reset my brain, I was forced to confront the truth.

The issue wasn't just my knees. A creeping anxiety had taken over my days. I was irritable, frustrated, and glum. There was nothing unfamiliar about any of this—and that was the real problem.

Over the past seven months, I poured all of myself into skiing. I'd created a series of goals that most people thought absurd, and I thought unlikely. The twenty tricks on my trick list and eighteen lines on my line list added up to the hardest physical challenge I'd ever attempted.

Yet I skied all those lines and learned most of those tricks. I also reached my initial goal of fifty days on snow and was one day shy of my stretch goal of eighty-six. If my knees would heal and I could get up to Mount Hood before the snow was gone, this too would be in the bag.

In other words, in every sense save sore knees, my Gnar Country experiment had been a success. But I'd done all this and come out the other side and little else changed. I was still me. I was the same damn me that started this quest. The same petty frustrations. The same shallow anxieties. The same flash temper. Why was I even surprised?

Most of us arrive in our fifties feeling that the cage has gotten smaller. What's actually shrunk is our mindset. We're in a prison of our own making. Once we discover we can keep on learning later in life, that mindset shifts. The cage vanishes.

This changes everything.

The most famous example of that change comes from Harvard psychologist Ellen Langer's "Counterclockwise Study," wherein a group of eighty-year-old men pretended—for five days straight—to be twenty years younger. This shifted their mindset and boosted their mood. The bigger change was physical. Participants who pretended to be younger—got younger. Their vision, hearing, memory, gait, posture, grip strength, joint flexibility, and manual dexterity all improved. The wildest result? Arthritis symptoms decreased so much that height and finger length increased, as participants could once again straighten out their joints.

Put differently: Aging is a fact of life. Old is a mindset.

And this was the real trouble with my knees. Unless I could heal them soon, I was headed back to mindset jail—where I was gonna be stuck with the same damn me for too damn long.

## NORTHERN NEVADA, JUNE 20, 2021

My knees got better. My knees got worse. My knees got better. My knees got worse. The uncertainty was making me crazy.

Today, I decided to solve the mystery. I would get an ultrasound. I would also get more aggressive with my treatment. But I knew what this meant and I didn't like it. It meant: snake oil and miracles cures, here I come.

Over the years, I've investigated an alphabet of treatments. Acupressure, acupuncture, bio-identical hormones, body work, chiropractic work, cold laser treatments, hot laser treatments, functional medicine, human growth hormone, microcurrent stimulation, microflora rebalancing, regenerative medicine, surgery, supplements—this list goes on.

Most of this stuff didn't work. Some stuff worked for a while. Some stuff worked too slowly. Some stuff worked only on Tuesdays. Some stuff worked only on Thursdays. Some stuff—well, it always felt like a crap shoot. As a result, I now have a single rule for assessing treatments: I like stuff that works like ice.

Ice works. It works for everyone, every time, and in the same way.

Reliable and repeatable the world over. The cold numbs pain and shrinks inflammation. No mystery science. No quantum fairy dust.

And this brings us to regenerative medicine, which is a catchall term for treatments that replace or regenerate cells, tissues, and organs. Treatments often begin with an ultrasound—to visualize the injury—followed by injections. What's being injected? That answer has changed over the years.

The first time I experimented with regenerative medicine was in the early 2000s while recovering from Lyme disease. I spent two years injecting everything from $B_{12}$ to bio-identical hormones to human growth hormone. The verdict: It wasn't ice.

A decade passed before I gave regenerative medicine another shot. A desperate need to heal a torn rotator cuff sent me back.

Over the years, I've endured four rotator cuff tears. With the first two, I went the traditional route—cortisone shots and surgery. Each took an exceptionally painful year and a half to heal. The third time, a friend suggested the next iteration of regenerative medicine, aka platelet-rich plasma (PRP) therapy.

The short version: icelike.

The long version: Your blood is drawn, then spun in a centrifuge. The spinning removes red blood cells and leaves behind platelets. Another centrifuge concentrates those platelets into a small amount of plasma. This platelet-rich plasma is then reinjected into the wound site. The serum contains high concentrations of the body's own tissue repair mechanisms—some thirty different growth factors and cytokines in total—that stimulate healing, no surgery required.

I was blown away. Before treatment, I couldn't lift my arm above my head. On the ultrasound, the rotator cuff tear was an inch long. Then I had two PRP treatments. Two weeks after the second, I was in the gym doing military presses with zero pain. Two weeks after that, a second ultrasound showed the tear had shrunk below a quarter inch.

These results were too good to be true. Afterward, I assumed I got lucky. Like, no way lightning would strike twice.

Then, a year later, I tore my other rotator cuff, tried PRP again, and got the same results. The verdict: icelike.

For my knees, PRP had been my plan. But using an ultrasound to guide injections is as much art as science, and my PRP doctor was back in New Mexico. As I was now in Nevada, I needed someone within driving distance with the same level of skill.

I called biohacker and fitness guru Ben Greenfield. Ben suggested I upgrade from PRP to the latest version of regenerative medicine: stem cells. He also suggested I visit Dr. Matt Cook in San Jose, California, claiming that Dr. Cook was the best ultrasound diagnostician he'd ever met, and a stem cell wizard.

Once again, apparently, I was off to see the wizard.

### CAMPBELL, CALIFORNIA, JUNE 23, 2021

I drove to Campbell, California, for my appointment with Dr. Cook. We started with an ultrasound. I was terrified. It was the moment of truth. I was about to find out if I needed surgery. I was about to find out if my quest to learn to park ski had come to a screeching halt. I thought I might puke.

The ultrasound showed that Fred and I had been correct in our diagnosis—I had an inflamed patellar tendon and two overstretched MCLs. I didn't puke. I didn't need surgery. My park skiing career had not come to a screeching halt.

Or so sayeth Dr. Cook.

Treatment was a series of injections. First, a placental matrix was shot into my knee. The matrix acts like a sponge, sucking up excess fluid, stabilizing the joint, and providing a scaffold for new tissue to build upon.

Next, exosomes are injected. Exosomes are what stem cells release.

They're nano-scale fat bubbles that contain messenger RNA to make proteins and microRNA to regulate gene expression. Essentially, once you get past the whiz-bang, exosomes repair the body and recruit other cells to help repair the body—or that's the theory.

In practice, after Dr. Cook finished treating my knees, he said six words that made absolutely zero sense: "Now get up and walk around."

"Get up and what?"

"Walk around," he repeated. "The pain should be gone, and the joint should feel stable. It'll only get better over the next nine months as the knees regenerate tissue, but you should feel the difference immediately."

"Bullshit," I said.

In my entire history of injury, no treatment has ever worked immediately. Sure, I've gotten some quick pain relief, but stability took months, often years. I hopped off the table and started walking.

"Holy shit," I said.

The pain was gone and my knees felt stable. I mean, really stable. Like way too stable for this to be the placebo effect. Like they hadn't felt this stable in decades.

Like ice.

### NORTHERN NEVADA, JUNE 28, 2021

I went to the gym for a workout, my first since the treatment. It was a crucial test. Since season's end, I haven't been able to lift heavy weights with my legs, nor attempt squats of any kind. Today, I lifted heavy weights and did squats.

Still stable. Still pain-free. Still ice.

### NORTHERN NEVADA, JULY 5, 2021

Over the next week, I kept at it, going back to the gym and doing weight vest hikes in between trips. The hikes were the bigger test, as they came closer to mimicking the dynamic forces involved in skiing.

A week ago, I wore a twelve-pound vest and took a thirty-minute hike. Three days ago, it was a twenty-pound vest and a ninety-minute hike. Two days ago, it was thirty pounds for an hour. Today, after waking up with no pain and plenty of stability, I called Ryan.

"Want to head up to Mount Hood?" I asked, when he answered.

"You think there's snow left?"

"Four hundred and fifty-five yards, to be exact," I replied. "At least according to their website."

"You think your knees are ready to ski?"

"I think there's four hundred and fifty-five yards of snow to answer that question."

### MOUNT HOOD, JULY 20, 2021, EIGHTY-SIXTH DAY ON SNOW

Mount Hood started running its chairlifts at 9:30. We were heading uphill by 9:31. I was jittery. It felt like my whole Gnar Country experiment was riding on what happened next. It felt like my future as a park skier was riding on what happened next. It felt like I hadn't been in flow for centuries.

Our plan was slow and steady. Ski over to the terrain park to dump our backpacks, then head to the main chairlift for some warm-up laps. But we had to ski through the park to get back to the lift . . .

"Showtime," said the voice in my head.

"Fuck off," I replied.

Still, I tried a tentative 50–50 across a dancer box. When that worked, I tried a small straight air off a roller. Then, I got ambitious and attempted a sliding spin 360 on the quarterpipe at the bottom. Finally, I breathed a sigh of relief.

Apparently, I can still ski in a straight line, jump into the air, and spin in circles without my knees collapsing.

Next, those four hundred and fifty-five yards of snow came to an abrupt end. We clicked out of our skis and hiked over an enormous bed of lava rock—like walking on the moon, like this used to be a glacier, like global warming right upside the head. On the other side, we found the next patch of snow, clicked back into our skis, and continued down the hill.

It took fifteen minutes to work our way back to the lift. It took three burners to get used to skiing at speed again. Those burners were taxing. My muscles were tired; my head was full of chatter.

By lap four, the chatter had subsided. I was leaning into my turns, full force on my knees, and everything holding steady.

By lap five, it was time for a real test. I unleashed Ryan. I knew he would go in search of air, linking jumps together into a flowy slopestyle course. I also figured, if I was chasing Ryan, all that air would arrive too quickly for the voice in my head to argue.

Ryan got after it. I got after Ryan.

There's a good bit of learning theory that says taking long pauses between practice sessions gives the brain time to move new information from short-term holding into long-term storage. As I didn't pause much during the season, I hadn't tested this hypothesis. Theoretically, it could mean I'm a better skier now than I was six weeks ago, despite not skiing at all.

Theory no more.

Over the next few laps, I threw all my tricks—nose grind, tail grind, shifty, tweaked safety, 180, nose butter 360, etc.—and without a lick of thought. Afterward, I calmed down. My knees felt fine, my hip flexor felt strong, my legs were still in ski shape, and I had those tricks on lock. My season hadn't been a hallucination. I was actually a very different skier than I'd been at its start. Most importantly, I had reached my stretch goal of eighty-six days, doubling my previous season's best.

"Impressive," said the voice in my head.

"Impressive?" I replied, grinning. "That's all you got?"

I was still grinning a few laps later, when we bumped into Jeremy Jones by the chairlift. Both a snowboarding legend and a longtime Squaw Valley local, Jeremy and I have known each other for years.

"What are you doing here?" he asked when I skied up.

"I came for the terrain park."

"I didn't know you knew how to park ski."

"I'm learning," I said. "Theoretically, it's supposed to be impossible for anyone over the age of thirty-five to get good. So, you know, I decided to see if I could get good in my fifties."

"Why? Wait, don't tell me, I already know—this is another one of your experiments."

I nodded.

Jeremy smiled. "So how many bones have you broken so far?"

As I said, Jeremy and me, we've known each other for years.

### GOVERNMENT CAMP, OREGON, JULY 21, 2021

I took the day off to rest. My knees felt fine. My muscles felt sore. A long walk, a little yoga, an Epsom salt bath, fingers crossed. . . .

### MOUNT HOOD, JULY 22, 2021, EIGHTY-SEVENTH DAY ON SNOW

We got a late start. The lifts opened at 9:30. We were heading uphill at a laggardly 9:32.

On the ride up, we were treated to the pro-show. The sixty-foot booter was in play. First, seven-time X Game medalist Sammy Carlson sent a bio 1080 safety to the moon, then two-time X Game Real Ski winner Phil Casabon, aka B-Dog, laid down a dirty misty 720 blunt, then Parker White going a million miles an hour, then Bella Bacon, maybe my favorite New Schooler, just seventeen years old and already a style queen.

At the top, we stashed our packs and got into our warm-up. The first lap was about moving my body. The second lap was about overcoming a fear barrier. Two days ago, while I'd thrown all my tricks, I hadn't wanted to ski backwards at speed. Maybe it was knee fear. Maybe it was the time off. Maybe it was time to solve this issue.

Thus, I slashed sideways, drifted backwards, and skied switch to the bottom of the run, no problem.

On the lift up, Ryan said: "That's the fastest I've ever seen you ski switch."

"Remember the first time you saw me ski switch?"

"I remember," said Ryan. "The bunny hill at Kirkwood. You couldn't turn. Couldn't even turn your head to see where you were going."

"It feels like a decade ago," I said.

Ryan did the math. "It was about six months."

The next three laps, we hunted for air. My knees continued to perform. My muscles started to feel better.

"Ready to work?" asked Ryan afterward.

I knew what he was asking. I knew what was next. I cranked up a mad dose of Slipknot and we headed toward the terrain park.

Then we skied into the park. I saw the mean rails, the pro-caliber crowd, the off-the-charts intimidation. I turned off Slipknot. Maybe I wasn't ready.

I stared at the medium jump line. Last year, I'd been too scared to try those jumps. In fact, I'd only attempted their kiddie line. I cleared zero jumps on that attempt and never came back for another go.

This year, there wasn't enough snow to build a kiddie line. Or a small line. So today—ready or not—revenge required the medium line.

Thus, I underjumped the first jump, barely reached the landing on the next, then mistimed my sliding spin 360 on the quarterpipe at the bottom and ended up throwing—whatever—at least that lap was over and done.

Next lap, we bumped into Eric Arnold at the top of the park. Seeing him again reminded me of his advice for jumping: Keep my weight pressed into the tongues of my boots. And that was how I cleared both of the hits on the medium line.

Eventually, we made our way back to the chairlift for some cool-down laps. I sort of remember skiing backwards very fast. I sort of remember getting low and trying to drag my hand across the snow. Sixteen laps later. . . .

We left the main run and made our way to a skinny track of snow near the edge of the glacier. Below us, a series of S-curves snaked between lava rock walls. Below the S-curves sat a strange halfpipe. The Sno-Cat drivers had been using the halfpipe dragon to harvest snow from the bottom half of this run to create the thin road that ran to the bottom of the mountain. What remained held the outline of a halfpipe, but instead of two parallel walls of snow, huge swatches had melted and mutated, leaving behind twin rows of giant sharks' teeth.

Yet the teeth looked skiable. They looked rowdy. It looked like you could air off some of these teeth and land on others.

I decided to go big and hope for the best, sailing a huge 180 into the pipe before I had time to change my mind. A snow grind got me to forward, then I jumped from one shark's tooth to another, then a 360 on one of the walls, then I learned a new trick—stalling atop a solitary tower of snow before hopping a 180 to the ground.

Then, finally, thankfully, after six weeks of wondering if this would ever happen again . . . hello, Ghost Dog.

### MOUNT HOOD, JULY 23, 2021, EIGHTY-EIGHTH DAY ON SNOW

Tired today, but excuses would not be tolerated. It was the last dance, the end of my season. Ryan and I got after it. Or tried. The slush was grabby, my legs were cooked, and the warm-up didn't help.

We went to the terrain park anyway.

I wanted to end the season with a steazy line through the medium jumps, something that would have been inconceivable last year. I thought about trying a series of 360s, but the lips were soft and nearly everyone was underjumping their spins. I settled for the old standards: a safety over the first hit, a shifty over the second, then a sliding spin 360 across the quarterpipe at the bottom.

Not bad. But not good.

I did it again. And again. Six laps later, I'd skied my single steazy line. Sure, it wasn't much compared to the steaze being laid down by other skiers in the park. Then again, I'm decades older than the other skiers in the park and, in the over-fifty category, I was killing the competition.

With the jumps out of the way, I wanted a single victory in the rails. None looked friendly. But there was a large propane tank that Ryan had been playing on all morning. Maybe I should try to rail-slide the propane tank. . . .

"Could we please," asked the voice in my head, "end the season without ass-crashing into another large metal object?"

Still, the tank had piqued my curiosity. I couldn't walk away now. That would leave unfinished business. That could violate the "no more shame" clause in my contract.

What if I tried to jump diagonally over the tank? That was a creative solution. That would sate my curiosity and protect my ass. Finally—an idea that made sense.

See the line, ski the line, that's the rule. But as soon as I launched, the voice in my head changed the rule.

"Tail-tap the tank," said the voice.

*Thwack!* with authority is what happened next. And take it from this old man—that shit never gets old.

Yet I was old enough to know not to push my luck, so goodbye to the terrain park and see you next year. Then Ryan and I zipped back to the chairlift and skied hot laps until neither of us could move. Then we got back on the lift because not being able to move did not seem like an excuse for not skiing.

From the lift, we noticed the sharks' teeth halfpipe again. Why not play a little fast geometry to end the season?

Fast geometry was a fairy tale. I was running on fumes. By the time we skied to the top of the S-curves, my legs were screaming. There was a crowd of snowboarders off to one side, but—as my legs were screaming—I didn't pay them much mind. Instead, as Ryan told me later, I was busy negotiating with my quadriceps.

"Come on quads," he heard me say as I skied by. "I need one more run. One more. That's all I'm asking. Just one more. . . . "

Negotiations must have broken down. Ryan heard me roar and saw me charge directly at the S-curves.

The melt-out had gotten worse. The first berm was a narrow bridge, the second berm required prayer. I whipped into the third berm at full Mach and, yikes—that berm was gone. There was lava rock where my turn used to be.

When in doubt, air it out.

I altered my trajectory and aimed straight for the berm, using it like a ramp and hoping—as I sailed over a bed of lava rock—there would be snow on the other side. There was snow. And more snowboarders . . . like, what the hell?

I landed beside an enormous pack of them. At least thirty snowboarders in total, including a couple lugging film cameras. Maybe some kind of movie shoot?

Then, I saw Ryan fly out of the S-curves and do the same double take I had just done. I caught his eye. We were both thinking the same thing—escape the crowd—beat the snowboarders into the pipe.

But the snowboarders didn't want to be beaten. The moment we dropped, they dropped. Thirty riders soaring into the sharks' teeth at once. It was a mob scene. But it was the last run of the season and I wanted it to count.

I jumped every hit in sight. I threw all my tricks. And so did everyone else.

My favorite moment came at the bottom of the pipe. I spun a final sliding spin 360 across the right wall and caught sight of a trio of snowboarders just behind me. Three women, all midair. Two sailed off the left wall, one leapt from the right; each was five feet above the ground, their long hair flying sideways out of their helmets. Then I spun to forward, and they were gone. Just another memory in a season packed with them.

Later, I realized what this memory actually meant: I blasted into a supercompetitive situation and instead of crumbling under the pressure, I used the added stress to drive focus and drop me into group flow.

File under: holy crap!

One reason I'd come to Gnar Country was to settle a score with "those fucking jocks"—aka to try to forgive those who had done me harm. There was no other choice. If I wanted to enjoy my later years, clearing this adult development hurdle meant moving past the dramas of my past. You can't fight your biology; you have to forgive your history.

Would learning how to park ski help me clear this threshold? Not a clue. But I had no other ideas.

In that halfpipe, I put myself into a very competitive situation, dropped into group flow, and contributed something to the mix. In that moment, I was one of those fucking jocks. And the strangest part: It was seriously fun—and why am I only learning this now?

When I think back on that memory, I always end up thinking the same thing: I learned a new language this season. Formally, it's the language of park skiing, but that's just a modern spin on an ancient tongue. This language of gravity. This language of courage. This language of possibility.

"The limits of my language are the limits of my world," said Ludwig Wittgenstein.

Yeah, he said it right.

# Chapter 10

### CAMPBELL, CALIFORNIA, AUGUST 6, 2021

I walked into Dr. Matt Cook's office at 11:00 a.m. This time, I packed a lunch and brought a book. There's nothing quick about regenerative medicine—that much I learned last time.

This time, Dr. Cook re-treated my knees and my shoulder. Five hours later, I walked out of his office again. My knees felt good. My shoulder was improving. And regenerative medicine felt like the final piece in the puzzle.

I designed my one-inch-at-a-time approach to promote skills acquisition and prevent catastrophic injury. Proof that it worked? I skied all the lines on my line list and learned most of the tricks on my trick list. Afterward, I didn't need surgery. A total success. Yet I hadn't anticipated the chronic damage that accompanied success. And this is why regenerative medicine mattered.

If I learned anything in Gnar Country it was that high levels of athletic performance are achievable later in life, yet success—aka the day-to-day pounding of success—takes a toll. I was pretty sure that better preseason training and more in-season recovery could alleviate most of that toll. Regenerative medicine seemed ready to handle the rest.

Yet that wasn't the reason I had come back to see Dr. Cook. After my adventure at Mount Hood, my body still felt great. I didn't need another round of injections to heal damage from last season. Instead, I wanted to stay ahead of the curve and prepare for the damage from next season.

In this way, my return to Dr. Cook's office was actually the ultimate proof of my Gnar Country success. When I started my season, park skiing had been an intense curiosity about an impossible goal. Well, been there, done that, and give me more. Now I was playing the infinite game—where the goal was to keep on playing.

This was why I went back to see Dr. Cook. When next winter rolls around, I wanted to be even stronger than I was last winter—aka Gnar Country, Part Two. Faster geometry. Game on.

## NORTHERN NEVADA, AUGUST 9, 2021

I woke up, wrote for four hours, took Kiko for a hike, ate breakfast, and went to the gym. In other words, I did the exact same thing I did all season, only my season was over. My return visit to Dr. Cook marked the beginning of my next season. Today's trip to the gym, in fact, was next season's first preseason workout.

When I got home from the gym, I pulled out the list of goals that defined my Gnar Country experiment and studied the results.

Eighteen months ago, I began with a single goal: "Enter next ski season a stronger skier than I ended last ski season." A few weeks later I made the decision to learn to park ski. Then, I added forty-six more goals to the list.

Eighteen of those goals were lines on my line list. Twenty were tricks on my trick list. One goal was to ski fifty days this season, which later became eighty-six days. Two goals were stuff to avoid: I didn't want any more surgery and I didn't want to end my season with any more shame or regret. Three goals were the steazy lines I wanted to ski through kiddie, small, and medium-size terrain parks, respectively.

The final three goals were more general: Be able to hold my own in a posse of hard-charging skiers. Be able to see an interesting feature on the hill and know enough tricks to do something interesting with that feature. Be a very different skier at the end of my ski season than I had been at the start.

Together, these forty-seven goals were my measuring stick. They were how I would judge the success or failure of my experiment. Today, I crossed the final item off my list. My season was over and I was a very different skier.

In total, I went forty-five out of forty-seven. The two unfinished goals were both tricks on my trick list. I had failed to learn a 360 surface swap on a dancer box and I had failed to learn a daffy butter.

The failure to surface swap was about opportunity. Neither Kirkwood nor Heavenly had a dancer box. The plan had been to delay learning this trick until the end of my season, when I moved to the bigger terrain parks at Mammoth and Mount Hood.

Two words: climate change.

The warmer temperatures increased snowmelt, and neither Mammoth nor Mount Hood could maintain their smaller terrain parks. By the time I arrived, they'd both disassembled their dancer boxes. I was left with no place to learn the trick.

My issue with the daffy butter was different. The trick combines a nose butter and a daffy—that is, hit a jump and kick one leg forward and one leg backwards while midair. Thus, my issue.

While I'd done nose butters all season, I'd never thrown a daffy before. I had no idea what kicking one leg forward and one leg backwards felt like or was supposed to feel like. My one-inch-at-a-time approach demanded building on existing motor patterns. But every time I considered the daffy butter, because I didn't know the foundational movements, there was never anything to build upon. I couldn't go one inch at a time. It turns out that my failure to daffy butter was actually a sign that I was following rule number one: Always follow the rules.

I could live with forty-five wins and two bits of unfinished business. I'd gone further than I'd expected to go. According to Ryan, I'd gone further than he expected me to go. The fact that I had skied thirty-eight days more than my original goal of fifty partially redeemed these misses. The fact that my knees were strong again and I could get into a trampoline park and learn the motions required for a daffy butter—so I could throw the trick on the first day of my next season—that was the real balm.

Yet these were minor details. The major detail: My peak-performance aging experiment had been a runaway success. Thus, on the off chance that you might be interested in conducting your own experiment, we need to translate the lessons I learned while skiing into all those nonskiing activities that people sometimes do—don't ask me why.

Before we get to these general principles, there's some stuff you should know. And the first thing to know: There's a lot of stuff you don't know.

My performance journal focused on what happened on the mountain. It omitted much of what happened everywhere else. So what happened?

More than a little.

For the past eighteen months, while I skied eighty-eight days, I also launched a book, edited a second, wrote a third, and almost finished writing a fourth. Additionally, I helped steer the Flow Research Collective through a pandemic, gave over two hundred speeches and interviews, led a half-dozen research initiatives, and managed to stay happily married throughout. Also, my dogs still like me.

All this matters for one big reason—being busy is not an excuse.

Too often, the siren song of adult responsibility is where our dreams go to die. We have an alphabet's worth of excuses: I can't do X because I'm already doing Y and Z.

Yet I ran my experiment while doing XYZ, ABC, and a bunch of other letters. I also went forty-five out of forty-seven and ended my season still hungry for more. More importantly, I brought nothing special to the table. I'm a bad athlete in a broken body with a busy schedule. And when my experiment started, I was sick with COVID and burned-out beyond belief. So whatever your adult reasons for not charging hard at big dreams, while I'm not negating the reality of those challenges, know that I too was up against it and still got it done.

And this brings us to the first general principle of peak-performance aging, and arguably the hardest principle to understand: Hard work works.

I started my season with a fairly ridiculous list of goals. I ended my season having accomplished almost all of them. What happened in between? The work happened. I had no special skills. There was no magic trick. I simply showed up every day and tried my hardest, while focused on a single clear goal: Ski sixteen laps.

Last season, whenever I got to the mountain, this was my aim: Ski sixteen laps. That was the hard work. Were they good laps? Bad laps? Did they feel good? Did I learn a new trick? Did I ski a new line? Didn't matter.

Sixteen laps was all that mattered.

This is one reason why wading through all that skiing was so important. If we didn't wade through all that skiing, no one—including myself—would have believed that I could achieve the impossible just sixteen laps at a time.

### PINE NUT MOUNTAINS, AUGUST 11, 2021

I put on a twenty-pound weight vest and took Kiko for a hike. Along the way, I started thinking about all the stuff I did last season that helped me succeed. This brought me right back to where I already was: my weight vest hikes.

The hikes are an example of hard work working. They were designed to get me physically ready for my Gnar Country adventure and mentally ready to maximize time in flow during that adventure. Over eighty-eight days, I almost always skied sixteen laps, often finished above twenty, and generally dropped into flow along the way.

More than being an example of hard work, the hikes are also an example of *smart* hard work. And this brings us back to *stacked protocols* and *multi-tool solutions*, the next two general principles we need to examine.

We'll go one at a time.

A stacked protocol is a way to do multiple things at once. Since I hike Kiko every day, adding a weight vest to that hike didn't add any extra time to my already busy schedule. Yet this one action allowed me to accomplish two tasks on my to-do list—hike Kiko and train for ski season—simultaneously.

Better still, the hikes are a well-established ritual, automatically executed. The only time I have to think about what I'm doing is the initial five minutes of the hike, when the extra exertion is noticeable and the voice in my head has plenty to say.

Afterward, because of my slow-and-steady approach to increasing vest weight and hike length, I barely noticed the extra exertion. Nor was I wiped out by the effort.

This was also why I wanted my mind to wander during my hike. Mind wandering was a sign of low-energy expenditure. It protected me from my natural inclination to push too hard. Since energy levels help determine the challenge-skills balance, pushing too hard on my morning hike threatened my chances of getting into flow during the rest of my day and my motivation to get after it again tomorrow.

Finally, these hikes check off an additional item on my daily to-do list. Flow is not a binary experience: in the zone or out of the zone. Flow is a four-stage cycle: struggle, release, flow, recovery. To maximize flow, we must maximize our ability to move through all four stages of this cycle. And this helps explain the order of my daily routine.

My day starts with a writing session. One of two things happens: The session is difficult and I stay in struggle, or there's a breakthrough and I drop into flow. If I'm in struggle, I want to follow this with a release activity, meaning a low-grade physical activity like hiking that, by slightly occupying the conscious mind, allows the brain to pass the problem over to the subconscious. If I'm in flow, this needs to be followed by a recovery activity, like a long walk in nature.

Because I follow my writing session with a weight vest hike through the mountains, I'm actually executing a triple stack of tasks—hike Kiko, train for ski season, and a release/recovery activity all at once.

Equally important, these hikes are also a *multi-tool solution*, meaning, a single tool that solves multiple problems at once. This matters because any adventure into Gnar Country involves three overlapping sets of challenges: healthy aging, peak-performance aging, and experiment-specific peak performance.

Healthy aging requires the daily application of the five Blue Zone keys to a long happy life: Move around a lot, de-stress regularly, have robust social ties, eat well, and try to live with passion, purpose, and regular access to flow.

Peak-performance aging has four mental components and five physical components that must also be trained regularly: motivation, learning, creativity, and flow on the mental side; strength, stamina, flexibility, balance, and agility on the physical side.

Next, there's experiment-specific peak-performance requirements. For me, this was everything from learning how to park ski to learning about the gear needed to park ski to learning how to recover from the damage done by park skiing, etc.

But one tool—weight vest hikes—solves multiple problems at once.

For starters, hiking builds leg strength, which peak-performance aging demands. In the elderly, leg strength is the single most important factor for longevity. Thigh muscle mass inversely correlates with mortality. Strong legs maintain physical functionality, boost overall health and levels of social activity, and protect against a common killer of older adults: slipping, falling, and breaking a bone. Beyond the bone break itself, the energy required to heal depletes the immune system, leaving us vulnerable to often fatal secondary infections like pneumonia.

Mentally, strong legs play an equally important role: they protect the brain. In a ten-year study with over three hundred twins, all over age fifty-five, researchers at Kings College London found a "strikingly protective relationship" between high leg strength and preserved mental abilities and brain structure. In fact, out of every health and lifestyle factor they examined, leg strength was the most important variable for maintaining cognitive function late in life.

Even better, hiking, like all exercise, is a potent de-stressor. Stress produces inflammation, which is the root of much that we call "aging." Insomnia, depression, anxiety, Alzheimer's, and cognitive decline all link to stress. And beyond these longevity challenges, there are peak-performance penalties. Stress weakens motivation, hampers learning, decreases creativity, and blocks flow.

Mental health is also why my hikes take place in the mountains. Time in nature lowers stress and improves mood. Plus, wide vistas enhance creativity, novel environments enhance learning, and both trigger flow. In this way, my hikes in nature are a single solution to four challenges associated with health and longevity and six challenges associated with peak performance.

But it's the weight vest that turns these hikes into a multi-tool solution worthy of a Gnar Country adventure. With a weight vest, a faster pace, and a longer hike, I'm also working on strength, stamina, balance, agility, and flexibility—or all the use it or lose it skills that we need to train as we age.

In total, my weight vest hikes solved fifteen different problems at once.

My goal, however, isn't to lobby for weight vest hikes. For me, because of who I am, how I like to live, and a bunch of other individual factors, these hikes are a perfect stacked protocol and multi-tool solution. Rather, my goal is to get you to understand that late-in-life peak performance is possible, but there's a lot to do. If you don't work smart, there's just too much to do for any of this to work.

Hard work works. Smart, hard work works better.

### NORTHERN NEVADA, AUGUST 14, 2021

Real life came roaring back. Most of Tahoe was on fire. The Dixie Fire was burning to the north. The Tamarack Fire was burning to the south. Today, the Caldor Fire broke out to the west. I saw the smoke plume rise above the Sierras. Some of my best friends are trees. The air was black with their ash.

### RENO, NEVADA, AUGUST 15, 2021

The fires were still burning. I needed a distraction. I needed some flow. Plus, it was time to get my off-season training program under way.

The solution to my needs: a trampoline park in Reno.

While I'd never been on a trampoline before, I thought it might help me solve problems I couldn't solve during ski season. Specifically, to practice grabs, spins, and flips—that is, tricks too dangerous to learn on snow. But all that was down the road.

First, I needed better air sense.

Proprioception is the ability to know where the body is in space and exactly what it's doing. In sports such as park skiing, proprioception is renamed "air sense," as this emphasizes the three-dimensional nature of the activity.

Since I wasn't a park skier, I started my experiment with little air sense. My skills improved over the first half of the winter, then hit a wall. After my T-boning incident, residual fear started messing with my ability to spin into the air during my nose butters. Why just my nose butters? I have no idea. I just knew I was stuck. But if I could use the trampoline park to increase my air sense and decrease my fear, then I could solve the problem before the winter arrived and start next season a stronger skier than I'd ended last season.

Anyway, that's the theory.

The reality: Ryan and I found the tramp park crowded with little kids. Once again, I'm at least two decades older than everyone else in the room. Whatever. Turns out that this reality is perfect for my reality.

My reality: I was scared of trampolining. I was afraid my body couldn't handle the impact and my knees would pay the price. I was worried that my attempt to lateralize—that is, move into a secondary activity (trampolining) to improve my primary activity (skiing)—wouldn't work and I'd never flip, spin, or grab with ease. Plus, I hate being bad at anything in public. I don't like crowds. I could go on but, by now, you know the drill.

Yet this extra pressure turned out to be useful, as my age was another reality. I was now fifty-four years old—which means "risk tolerance" is another use it or lose it skill that I have to regularly train.

Risk tolerance declines over time, starting in our midtwenties and dropping steadily until our midsixties, when it flattens out again. In any Gnar Country adventure, this is problematic for a number of reasons.

Gnar Country is about chasing down big dreams before it's too late to chase down those dreams. Risk is built into that effort. Risk is also a flow trigger, and managing risk is critical to tuning the challenge-skills balance, which is flow's most important trigger. Novelty and unpredictability are also flow triggers, yet our tolerance for novelty and unpredictability is coupled to our tolerance for risk.

A bigger issue: The superpowers that come with aging are not guaranteed. If we want the decrease in ego and increase in intelligence, empathy, and creativity that become possible in our fifties, we have to clear the three hurdles of adult development—all of which require risk taking.

By age thirty, we need to solve the crisis of identity. By forty, we need to find "match quality," which is an alignment between who we are—our beliefs, values, skills, and strengths—and what we do for a living. By fifty, we need to put down resentments, and forgive those who have done us wrong. Yet even if we pass through these gateways, we can't access our superpowers if we don't maintain our risk tolerance on the other side.

Fear blocks empathy and decreases creativity. Worse, it creates stress, which causes health problems, which further lowers our appetite for risk and further increases our stress. If we're not careful, we can ride this feedback loop of risk aversion right into an early grave—which is ironic, as the reason most of us stop taking risks is out of fear of that early grave.

In short, I'd come to the tramp park to increase my risk tolerance, develop some air sense, improve my park skiing skills, and, hopefully, find more flow. It's another multi-tool solution—or, once again, that's the theory.

My reality unfolded in stages. The first stage was discovery. I discovered that bouncing on a trampoline used a whole bunch of muscles I didn't even know existed. I discovered awkwardness, exhaustion, and a mistimed bounce—which sent me chin-first into the trampoline and, fuck, did that one hurt.

The second stage involved tuning the challenge-skills balance—that is, running away from the big, mean Olympic trampoline and heading toward the baby tramp that bounces into a foam pit.

On my way over, the voice in my head offered a suggestion: "Front flip."

*God help me . . .* was my next thought. I was still sore from the Olympic tramp, had never bounced into a foam pit before, and twenty-five years have passed since my last attempt at a front flip. No way did I want to attempt one now. So I did.

Six attempts later, I flipped onto my feet. Set goal, meet goal—hello, neurochemistry.

Next, I started trying to tweak my flip angle. Instead of going ass over teakettle, I practiced going ass sideways over teakettle. The results weren't much to look at, but I'd never thrown an off-axis flip before and—hello, flow.

Now it was time to return to the Olympic trampoline and get some revenge. And this brings us to another general principle worth reiterating: Flow first, risk second.

You need to train up risk tolerances for peak-performance aging. Yet as Glen Plake taught me, using risk, especially physical risk, as a flow trigger is dumb and dangerous. If you have a choice, you want to be in flow and already performing at your best before attempting the dumb and dangerous.

This time around, me and the Olympic tramp got along just fine. Better than fine. I figured out the basic motions for a daffy in less than five minutes. Then I went grab crazy. First, a mute grab; then a Japan grab; finally, a stink bug grab. What were all these grabs? Doesn't matter. All that mattered: I'd never thrown any of them before.

I was making progress. I was still in the game.

### NORTHERN NEVADA, AUGUST 16, 2021

Sore. Very sore. Hurts. Hurts to move. Hurts to think about moving. Instead, I stayed in bed all day thinking about all the ski tricks I can learn once the soreness goes away and I can head back to the trampoline park.

It's a sickness, I know.

## NORTHERN NEVADA, AUGUST 17, 2021

Still sore, still hurts to move, still can't stop thinking about all the tricks I want to learn on my next trip to the tramp park.

And this brings us to Gene Cohen.

If I was keeping office hours, I would say that Gene Cohen is the godfather of peak-performance aging. In the 1970s, Cohen pioneered the field of geriatric psychiatry, which is the field devoted to the mental health of the elderly. If you're wondering why it took us until the 1970s to realize that the mental health of the elderly was a topic worthy of scientific study—well, good question.

Before the 1970s, scientists thought of aging as an inevitable long, slow rot. Everyone agreed: Depression, loneliness, and cognitive decline were built into the rot, and there was nothing we could do about these facts.

Not so fast, said Cohen.

In 1973, as a young psychiatrist working at the National Institute of Mental Health, Cohen thought there was more to the story. He also knew that the boomer generation was both the largest generation in American history and starting to get on in years. If mental health outcomes for older adults didn't improve, the boomer drain on public resources would be immense. Thus, Cohen lobbied his employer for the creation of a research institute devoted to the then radical idea of "successful aging."

It worked.

Cohen became the first head of the NIMH's Center on Aging, a position he held for nearly fifteen years, before becoming the director of the National Institute on Aging for another five years. In both of these positions, Cohen oversaw two of the largest and longest studies on aging ever conducted.

What did he learn?

Rather than a long, slow rot, Cohen discovered, our later years are a new stage of adult development, where three profound and positive changes take place in the brain. First, certain genes activate only by experience, which means the brain remodels itself over time, adding depth and wisdom to our personalities as we enter our later years. Second, the brain learns to recruit regions underutilized in our earlier years, and this can help compensate for the cognitive decline that comes with age. Third, the brain's information processing capacities reach their greatest density and height between ages sixty and eighty, allowing the two hemispheres of the brain to work together like never before.

These neurobiological changes unlock three types of thinking that are mostly inaccessible before our fifties. More importantly, all three types of thinking continue to improve with age, as long as we continue to cultivate creativity—which is what's required to train up these thinking styles.

1. **Relativistic Thinking:** We learn to better synthesize disparate views. We learn that there are few absolute truths, mostly relative truths, and that black-and-white thinking is a folly of youth.

2. **Nondualistic Thinking:** We learn to consider opposing views without judgment. We learn to see both sides of the same coin. We learn empathy.

3. **Systematic Thinking:** We learn to think big picture. We learn how to see the forest through the trees. We learn to think divergently.

Cohen's work changed the conversation around peak-performance aging. It's no longer a question of what we can do in spite of our age. Now it's about what we can do because of our age. Or, as Cohen explains in *The Mature Mind*, his classic book on the subject: "Dozens of new [neurological] findings are overturning the notion that 'you can't teach an old dog new tricks.' It turns out that not only can an old dog learn well, they are actually better at many types of intellectual tasks than young dogs."

In other words, if my Gnar Country experiment proved anything, it proved that Gene Cohen was right.

### NORTHERN NEVADA, AUGUST 20, 2021

I reread my performance journal. I needed proof.

Even though I spent my year attacking lingering issues of the "courage, shame, and those fucking jocks" variety, I didn't feel braver.

But what did I find rereading my journal? Proof of bravery.

Nearly every day of my season, I'd done something that terrified me. Every day, I edged out of my comfort zone, pushed on the challenge-skills balance, and was generally successful in my efforts. So I ask again, why don't I feel braver?

I didn't know, so I called Dr. Sarah Sarkis. A brilliant clinical psychologist who has worked with the Flow Research Collective since its inception, Sarah's also the person I call in times of confusion.

"I'm confused," I said when she answered.

"Tell me more," replied Sarah.

I told her more. Sarah already knew about my Gnar Country experiment. I told her that the experiment was over and it had been a runaway success. But I didn't feel any braver. I told her I had reread my performance journal and found proof of bravery on every page—but I still didn't feel any braver.

"Interesting," said Sarah. "Tell me more."

When Sarah says "interesting," she means it like an entomologist staring at a bug—but whatever, I told her more.

"I don't feel any different," I said. "I love skiing. I just skied for seven months straight—which is a miracle. I accomplished my goals.

The experiment was a success. But I still feel like the exact same me that started the experiment. Also, since I stopped skiing, all this crazy shit's been happening. . . ."

"I love crazy shit," said Sarah, almost giddy. "Tell me more."

I told her that my summer has been déjà-vicious. "Every challenge that drove me mad enough to start my Gnar Country experiment has reappeared," I said. "I mean, every single challenge. One earthquake, three forest fires, a COVID surge, trouble at work, trouble at home, my body broke down, my burnout returned, couldn't write, totally locked out of flow, and here's the weird part—instead of freaking out, I just deal, and there's like no emotional reaction whatsoever. None."

Sarah started laughing.

"You mean," she said, "you feel calm in the face of a crisis. You do know that's called 'courage,' right? Being calm in the face of crisis."

"Dunno," I said. "Never felt anything like it before. The shit's freaking me out. . . ."

## RENO, NEVADA, AUGUST 28, 2021

Return trip to the trampoline park, and what did Margaret Atwood say? "Everybody I know is an adult. Me, I'm just in disguise."

Yup, that's it exactly.

## NORTHERN NEVADA, AUGUST 29, 2021

Yesterday, Ryan and I tested out the tramp skis for the first time—that is, skis with rubber bottoms designed to work on trampolines. Yesterday, while wearing those skis, I nailed my first mute grab, blunt grab, and stale fish grab. Yesterday, I drove home from the tramp park basking in the glory of success.

Today, I can't move. Today, it feels like someone beat on my lower back with a baseball bat. Today, once again, I'm negotiating with success.

I called big wave surfing legend Laird Hamilton to discuss the negotiation. "I spent my ski season doing the gnarliest stuff I've ever done," I said. "And like ninety-nine percent of the time—it worked."

"So what's the problem?" Laird asked.

"Success. I got so addicted to the feeling of success that I kept pushing. I fucked up my knees because, every time I did something I thought was impossible, it was immediately followed by that tug. 'OK, what else is possible?' But 'what else is possible?' is like the most addictive drug I've ever encountered."

"Yes," said Laird.

"Worse," I continued. "A lot of the stuff I did—even though I pulled it off—it was so scary that I ended up with PTSD. I've never had this kind of success before. I had no idea that success can lead to PTSD."

"Yes," said Laird.

"And today, my back feels like hell. Today, I realized that if I want to trampoline on a regular basis, I'm going to have to take my fitness to a whole new level. My experiment is supposed to be over. Yet I'm still negotiating with success."

"Yes," said Laird.

And this brings us to the next general principle of peak-performance aging we need to discuss—be prepared to negotiate with success.

## SIERRA NEVADA MOUNTAINS, SEPTEMBER 1, 2021

Real life collided with ski dreams. The Caldor Fire reached the Sierra-at-Tahoe Ski Resort and was threatening both Kirkwood and Heavenly. All three resorts turned their snow-making guns against the flames. The fire crews worked nonstop. The people were evacuated. The plants and animals were left behind.

I stood at my window and stared at noon skies blacker than midnight. Like climate change was eating my soul. Like nuclear winter.

And all the peak-performance training in the world couldn't stop me from sobbing like a baby.

### RENO, NEVADA, SEPTEMBER 8, 2021

Return trip to the bouncy house. It was my third visit. Theoretically, I now had enough of a trampoline foundation to begin to deploy my one-inch-at-a-time approach.

Today's inch: a 180 with a safety grab.

I'd been thinking about this trick for weeks. I think it requires five steps: jump into air, start to spin, latch the grab, spot my landing, stomp my landing. I think, if I can learn this trick before next season, that would be a stunning improvement during the off-season.

I also think, the map is not the territory. The map of Gnar Country shows a fearsome land. What the map doesn't show? The new learning theories beneath my one-inch-at-a-time approach. The superpowers that come with aging. The turbo boost that comes with flow. The increasing potential of regenerative medicine. Most importantly, the map doesn't show the future.

When I started my experiment, I asked myself a question I always ask at the start of such experiments: "If I learn this thing, how will my life be different in five years?"

I mention this now, because when I first wrote out my goals—the tricks on my trick list and the lines on my line list—five years was my honest estimate for how long I thought it would take to reach them. In other words, I resigned myself to the fact that I might be sixty years old by the time I learned to park ski.

But skiing in flow is the best I feel on this planet. And if I learned how to park ski, I would have a million more entrances into flow than were available before—which is my version of a retirement plan. Thus, I ran the experiment anyway.

Turns out, the experiment took eighteen months. Turns out, on my third trip to the bouncy house, I learned how to throw a 180 with a safety grab. Turns out, the map is not the territory.

I left the trampoline park obsessing about the possibilities for my new trick. Once I got home, I knew what to do next. I went to my office, pulled out a piece of paper and wrote, "2022 Trick List" across the top of the page. On the first line, I wrote, "180 with a safety grab." Next, I added the three additional grabs I'd learned: stink bug grab, Japan grab, and a mute grab. Then I paused.

Earlier today, I'd attempted my very first "single truck driver" grab—kicking my right ski in front of my face, then grabbing it near the tip. Scary, but I nailed it on my third try.

Now I added single truck driver to my list as well. Then, I thought about trying to hold that single truck driver while also executing a 180—a seriously steazy trick that seriously terrified me.

Before I had time to change my mind, I wrote 180 with a single truck driver on my list. Sure, this goal might take me five years to achieve. Sure, I'd be nearly sixty years old by the time I got it done. But a sixty-year-old man throwing a 180 with a single truck driver? What's more punk rock than that?

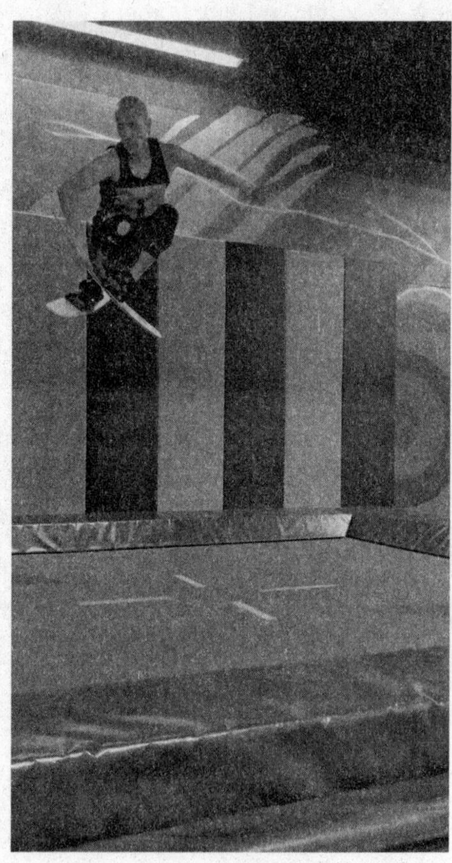

**From left to right:** Single truck driver grab, mute grab, Japan grab, stink bug grab, and safety grab.

SKIADDICTION

# Appendix: The Rules

In Chapter Ten of this book, we reviewed a series of general principles that guided my Gnar Country adventure. All of them were important to my success. All of them were based on lessons learned along the way. Yet, over the course of that journey, I also followed an ironclad set of rules—many of which predate these principles.

There are thirteen in total. Originally, these rules were meant to counteract counterproductive tendencies and/or to provide guidelines for action in times of extreme stress. They were each designed to keep me out of the hospital and free of regret, usually created after the opposite decision sent me to the hospital and/or filled me with regret.

In other words, these rules were lessons learned the hard way.

Most will be familiar. Many were named as we went along. A few were discussed, but not formally called rules. Whatever the case, like the general principles, the point here is not to blindly adopt my rules. The point is to figure out your own counterproductive tendencies and stress-addled missteps, then use that information to create a set of ironclad rules to protect you from yourself.

Here's my full list.

## RULE ONE: ALWAYS FOLLOW THE RULES

The rules are unbreakable. For me, this lack of options is the best option. I want to take choice off the table. I want to remove emotions from the equation. Unless my emotions are hindering my performance (see Rule Three), how I feel in the moment matters far less than how I feel later, looking back on this moment, with the wisdom of hindsight.

The moment lasts a moment; hindsight lasts a lifetime.

Thus, how I feel in hindsight, looking back on my decisions, influences the overall quality of my life far more than how I felt in that moment. This is also why I "work for the boss." The boss is the version of myself who already has hindsight, who always has my long-term best interests at heart, and who uses both to create the rules.

In normal life, the rules are helpful. In Gnar Country, they're a lifesaver. In high-risk situations, the brain becomes logical and linear. Short-term thinking dominates. Creativity decreases. Choices are limited. And this does not always produce the best outcomes. In my case, vertigo makes this worse. Thus, this rule prioritizes incremental progress, protects against bad decisions, and guards against indecision.

Moreover, if you can't keep your word to yourself, then you can't trust yourself. Trust in oneself—aka self-confidence—is reflected in everything from perceptual bias to levels of intrinsic motivation to where the brain sets the challenge-skills balance. If you continually keep your promises to yourself, self-confidence never wavers, motivation becomes momentum, and flow becomes a regular occurrence. The result: You go farther faster and with a lot less fuss.

Most importantly, this rule *always* applies to goal setting. If you consistently break your word to yourself, once you set a goal, your brain immediately starts hunting for an easy way out. Meaning, if you don't keep your word to yourself, your brain moves into "I quit" mode long before you've even gotten in the game.

The opposite is also true.

If you never break your word to yourself, then once something becomes a goal—that is, once you say it aloud or add it to a goal list—then you never stop working toward that goal. The brain is a goal-directed, puzzle-solving machine. If it knows that quitting is not an option, then it never stops trying to figure out how to reach that goal, which is usually what's required to reach that goal.

## RULE TWO: IF RYAN CAN DO IT, SO CAN I, THUS I MUST, AKA FIND A TRAINING PARTNER YOU CAN TRUST

This rule is about fighting fear and maximizing progress. It takes decision making out of my hands during scary situations.

If Ryan does something and I don't, historically, I regret my decision later. I also end up burning lots of calories trying to build up the grit and courage to do what should have already been done and what—the vast majority of the time—I end up doing eventually.

This rule is a trade-off between unpleasant options. In doing what Ryan does, I'm choosing to fight a short battle with extreme fear, rather than a long battle with anxiety and shame.

Additionally, since Ryan is a slightly better skier than me, following his line is a way of automatically pushing myself to the edge of my abilities. This forces me into the challenge-skills sweet spot, amping up focus and flow, and ultimately shortening my path to mastery.

Yet this rule demands the right training partner.

After nearly a decade of skiing together, Ryan and I trust each other completely. Why? Because we consistently go out of our way to keep each other safe. When I get hit with vertigo, Ryan tends to notice the symptoms before I do. Often, he reminds me that the smarter option is to back off and come back later—even if this means he doesn't get to ski a line he's dying to ski.

Equally crucial, Ryan and I like to learn in the same way—meaning, in private, at high speeds, and very far away from other people. We have similar ski styles, warm-up routines, terrain preferences, fitness levels, ski abilities, body types, tolerance for suffering, and favorite flow triggers.

This last bit is especially key.

Ryan and I always start seeking flow by slowly upping the challenge-skills balance: skiing each lap slightly faster than the last. Next, we add in pattern recognition, attempting to creatively interpret terrain features. Then, we turn to the novelty and unpredictability that comes from exploring new zones. Finally, we deploy risk, which only becomes a real option once we're both in flow. This overlap in trigger preferences means we're never at odds with one another in how we want to approach the mountain.

Additionally, Ryan and I have worked hard to master and deploy the triggers for group flow, including familiarity (I know his twitches, he knows mine), equal participation (we take turns being the leader and the follower), good communication (both verbal and nonverbal), shared risk (we're taking the same chances), shared goals (we both want to be great park skiers), and so forth.

The result—put us together and flow is the usual result.

Finally, a considerable pile of data shows that maintaining healthy interpersonal relationships is fundamental to longevity. On a personal note: Well, shit, man.

Most of the time, I like animals more than people. I'm also an introvert and a writer, meaning, both my job and my sanity demand solitude.

Yet because I ski with Ryan, I get quality bonding time without having to fight off my natural isolationist tendencies or alter my schedule—something, historically, I'm unwilling to do for the sake of quality bonding time.

Also because "group flow" is such a powerful bonding experience, I can maintain my healthy relationships in less time than is normally required—which, truthfully, is the only way I could check this box.

## RULE THREE: IF RYAN CAN DO IT, SO CAN I—UNLESS THREE CRITERIA ARE PRESENT, AKA BE AGGRESSIVE BUT NOT STUPID

One: If I am feeling too much fear, if the emotion has begun interfering with my performance, especially if vertigo is involved, back off and try again later.

Two: Don't execute if exhausted. Once I start underjumping hits and making weak-ass turns—two clear signs of exhaustion—I am done playing hero for the day.

Three: My goal is slow-and-steady improvement. In the park, Ryan is a much better athlete. He can do a bunch of stuff I can't—like sending 540s to the moon off huge kickers—so, for now, copying his lunar explorations remains off limits.

Chasing Ryan ensures progress. These three exemptions ensure that progress does not lead to catastrophic injury. I'll also say that learning to identify the line between "fear" and "too much fear" was one of the more difficult things I had to do all season—which explains the next rule.

## RULE FOUR: SEE THE LINE, SKI THE LINE, AKA TRUST YOUR NEUROCHEMISTRY

This rule is about listening to internal signals. It's the difference, at least for me, between fear and too much fear.

When I "see the line," I'm seeing a puzzle snap together and feeling the neurochemical thrill that accompanies pattern recognition. That's the signal I'm looking for—it means it's go time.

The reason to ski the line immediately is that the dopamine generated by pattern recognition doesn't last for long. While it's around, it amplifies everything from courage to focus to fast-twitch muscle response. When that window is open, I have the best opportunity for success—so it's go-go-go before it's gone-gone-gone.

There are also variations on this theme. A slightly different version is, when I'm in flow, and the voice in my head offers a suggestion, take it immediately or risk getting knocked out of flow. Another version arises during games like Chase the Rabbit—if I watch Ryan do a move and my brain recognizes the pattern and I get the go signal, then I know I can execute the same pattern and should do so, immediately.

## RULE FIVE: FIGURE OUT WHO YOU ARE AND HOW YOU LIKE TO LEARN, AND DON'T WAVER

Be yourself, especially when it comes to learning styles. I'm an introvert with a long history of injury and a slow-and-steady approach to progress.

As an introvert, I want to ski out of sight, and far away from other people. My injury history means I favor style and substance over "Just send it, brah." And my slow-and-steady approach means that while I want to push myself on a daily basis, I want to be the one who decides when and how far.

This is how I learn best. Trying to learn any other way is absurd. Every chance I get to go skiing is a gift. Why would I waste this gift by putting myself in a position not to maximize my full potential?

Moreover, peak-performance aging demands that I keep trying to maximize my potential. "Older persons who pursue [challenging] activities in which they experience a sense of control and mastery are healthier both physically and mentally than those who do not," Gene Cohen explains in *The Mature Mind*.

This health boost comes from the tight relationship between the nervous system and the immune system. A sense of control and the feeling of mastery are two of the most positive emotions available. Positive emotions increase the production of both T cells, the white blood cells that fight off disease, and natural killer cells, which are the larger white blood cells that target tumors and other sick cells.

This is also why reliable access to flow matters as we age. The state produces feelings of mastery and control. It also produces feelings of happiness, meaning, and purpose. Thus, reliable access to flow isn't just how we create a life worth living—it's how we create *a whole lot more life* worth living.

## RULE SIX: RYAN'S RULE: NEVER STOP AT THE TOP, AKA HESITATION IS NEVER YOUR FRIEND

Hesitation opens the door to fear. Anxiety produces adrenaline, which burns a lot of calories and saps energy levels. The more tired I am, the worse I ski, both increasing my chance of injury and decreasing my chance at progress—which is the first reason never to stop at the top.

The second reason is that stopping at the top also gives the voice in my head time to voice its opinion. This alone can tip me out of the challenge-skills sweet spot, shatter focus, and block flow.

The third reason: Not stopping at the top means skiing at higher speeds on the way down. When the brain recognizes affordances—that is, opportunities for action—it takes qualities like momentum and rhythm into account. A body moving at speed can choose from a much larger catalogue of possible actions and, by extension, has a much greater chance of dropping into flow as a result.

## RULE SEVEN: IF YOU CAN'T KEEP UP, I CAN'T SKI WITH YOU, AKA FLOW FIRST, FRIENDS SECOND

If you can't keep up, I can't ski with you. This isn't personal. It's psychological. Skiing fast is what I do for sanity. Skiing fast means no time for thought, which means no chance for the voice in my head to get its hooks into my brain. It's a mental health concern.

Skiing fast also increases my chances of flow, which is always my first priority. At the Flow Research Collective, we often stress the importance of one's "primary flow activity," or the thing you've done most in your life that is most likely to drop you into the zone. For me, this is skiing. For you, it could be reading or coding or hiking or gardening or surfing or skydiving or surfing while skydiving, aka sky surfing.

Whatever the case, as we age, adult responsibility tends to get in the way of our primary flow activity. We set aside "childish things." We stop doing the activities that provided us with the most flow. It's a huge mistake—especially later in life.

Flow fights anxiety. As we enter the state, stress hormones are flushed out of our system and replaced by performance-enhancing, feel-good neurochemicals. Anxiety produces inflammation, which accelerates aging. Meanwhile, positive emotions stimulate the immune system and slow aging.

The bigger boon may be psychological.

For me, regular access to flow serves the same function as therapy—it calms me down, keeps me out of my head, and widens my perspective. The result? I'm a better version of myself. I'm more productive, creative, calm, and effective. Health, happiness, and overall well-being increase. My relationships are better. My life feels more meaningful.

So, yeah, if you can't keep up, I won't wait. Just like, if I can't keep up, I don't want you to wait. Otherwise, we're just standing in each other's way and blocking flow—and that's just not how friends treat one another.

## RULE EIGHT: EMOTIONS ARE TOOLS

I did a lot of roaring in this book. I used these roars to tap into anger and generate testosterone. I used the hormone to increase energy, amp up focus, and drive flow.

Two things are important here.

First, despite its masculine reputation, testosterone works the same for all genders. When humans get angry, the body generates the hormone to prepare for the fight to come. This is why martial artists regardless of gender deploy a *kiai* during battle, which is the Japanese word for a "short shout uttered when performing an attacking move."

Second, it's not just testosterone that works this way. All of the performance-enhancing neurochemicals named in this book—dopamine, norepinephrine, serotonin, oxytocin, endorphins—tie to emotions.

In Mammoth, for example, I used curiosity to fight back against vertigo. When I was dangling a hundred feet in the air, I tried to ignore the height by scouring the mountain for interesting lines to ski. Just by hunting those lines—that is, by cultivating curiosity—I caused my brain to release dopamine, which decreased my vertigo by increasing everything from focus and flow to courage and joy.

From a peak-performance perspective, emotions are tools. They're a fast way to generate neurochemistry, which offers a faster path toward mastery. The roaring, though—that's optional.

## RULE NINE: NEVER WASTE A FLOW STATE

Flow amplifies our ability to do, as well as our ability to remember what we did. Why? So we can do it again later. This means, when in flow, I always want to push myself, take chances, and learn new skills.

For similar reasons, when in flow, I protect the state against all intruders. To remain in the zone, the prefrontal cortex must stay offline. Anything that wakes up the voice in my head is to be avoided at all costs.

Thus, when I ski, I turn off my phone. No alerts. No email. No texts. It's also why I hate talking "business" or "writing" or "flow" on the chairlift. My goal is to focus on skiing and to block out anything that breaks that focus.

## RULE TEN: TRAIN LIKE A PRO, RECOVER LIKE A PRO

If you want to kick ass until you kick the bucket, you have to train for old age like a professional athlete. Studies show that with proper training we can retain seventy percent of our physical abilities until even very late in life. Better still, since the brain figures out how to compensate for some of what is lost, we can perform as though we've retained an even greater percentage of those skills.

But it's use it or lose it across the board.

There are five major categories of athletic performance: strength, stamina, agility, balance, and flexibility. Old age means that all five demand constant attention. It means you can't skip steps and you can't coast. Once you reach fifty, if you're not moving forward, you're sliding backwards.

Additionally, it's not enough to train like a pro; we also have to recover like a pro. Welcome to Gnar Country—where nap time is mandatory.

For me, recovery means high-quality nutrition, constant hydration, an abundance of anti-inflammatory supplements, tons of sleep, and zero booze. It also requires a daily hike, one to two daily yoga sessions, and a daily Epsom salt bath and/or infrared sauna. Finally, it also means availing myself of the entire regenerative medicine toolkit.

The active recovery portion of this protocol—that is, long walks and yoga—is especially important. One of my major Gnar Country discoveries was how much my body lied about fitness readiness. There were many days when I woke up exhausted, but I had to walk Kiko and why not stretch a little afterward and, nine times out of ten, I felt better, drove to the mountain, and radically exceeded my expectations.

## **RULE ELEVEN:** GRIT IS THE LAST RESORT

Grit is a limited resource. On a daily basis, grit links to willpower, which declines over time. Our willpower is highest in the morning, when we're fresh from sleep. It dwindles during the day, as we expend energy. Since any adventure into Gnar Country will demand grit, we want to conserve our fuel for when we need it most, not waste it casually when other options are available.

Similarly, over longer time scales, always relying on grit is a recipe for burnout. Since burnout is a complete derailment of progress, in Gnar Country, we don't tough it out unless we have to tough it out.

When faced with a challenge, instead of trying to be gritty, I fight back with the big five intrinsic motivators—curiosity, passion, purpose, autonomy, and mastery—though this is easier explained by example.

Take mastery. Using mastery as a motivator means tuning the challenge-skills balance. For example, if I ski a couple of big scary lines, instead of trying to push through my fear and keep going, I lateralize.

Now I head to the kiddie terrain park—where there's less fear and an entirely different kind of challenge.

Then I push myself in the park until I scare myself again, then lateralize once more. Now it's into the woods for high-speed pinball or onto the groomers for high-speed burners.

Grit's finite nature is also why Ryan and I spent so much time playing games like Chase the Rabbit. When playing, we're in a failure-free zone. The pressure is off. The fun level is high. Instead of trying to get to work—the deliberate practice approach to progress—we follow our curiosity and explore our passion, or what might best be described as *deliberate play*.

Since deliberate play produces more flow than deliberate practice and flow boosts learning rates, deliberate play actually takes me farther faster and without depleting my limited grit supplies.

In short: Hard work works; smart work works better; smart play works best.

## RULE TWELVE: GET PAID TO GROW OLD

Two decades ago, while researching rare and expensive musical instruments, I learned that Antonio Stradivarius crafted nearly half of the most valuable instruments in history. This caught my attention. But then I discovered an even stranger fact—two of Stradivarius's most famous violins were built when he was ninety-two years old.

I was baffled. Everything I then knew about aging—aka the long, slow rot theory—said that Stradivarius's feat was impossible. If physical skills decline significantly over time, including violin-making mandatories like fast-twitch muscle response and fine-motor control, there should be no way a ninety-two-year-old man made two of the rarest musical instruments in history. Even stranger: Stradivarius pulled this off in 1736—that is, long before the advent of modern medicine.

The mystery of Stradivarius stuck with me. In trying to solve the puzzle, my research led me into the then-nascent field of peak-performance aging. There, I discovered that fast-twitch muscle response and fine-motor control are use it or lose it skills. Since Stradivarius never stopped making musical instruments over the course of his lifetime—he made over a thousand in total—these skills didn't atrophy. Or, if they did atrophy, perhaps his brain figured out how to compensate for the loss.

But how could his brain compensate for that loss?

NYU neuropsychologist Elkhonon Goldberg helped answer this question. Goldberg discovered that expertise makes us resistant to cognitive decline. More specifically, he found that the brain's pattern recognition and pattern execution abilities persist throughout our lifetime, even in the face of significant neurological insult.

After reading Goldberg's work, I started to wonder if there might be other skills that persist over our lifetime—which is really where this book began. This was also how I bumped into Gene Cohen's discovery of the superpowers of aging. Taken together, these three bodies of research—the use it or lose it nature of our physical skills, the persistence of pattern recognition and pattern execution abilities over time, and the increase in intelligence, creativity, empathy, and wisdom that comes with age—solved the mystery of Stradivarius.

Under other circumstances, this might have been the end of my inquiry. But during this same period, I was also traveling around the globe, meeting with hundreds of CEOs, and helping them harness the power of disruptive technology and flow science to radically improve organizational performance. Along the way, there were lots of discussions about hiring, training, and what skills mattered most in the twenty-first century. Nearly every CEO said the same thing—creativity and empathy were the attributes they needed most in their companies and had the greatest difficultly hiring for or training in.

These CEOs needed creativity to drive innovation and help their organizations keep pace with the accelerating rate of change in the world. They needed empathy because company success depended upon team performance, which required collaboration and cooperation and both were tricky without empathy. They also knew that customer-centric thinking was a mantra for twenty-first-century business, and, without empathetic employees, no one had the ability to think like their customers.

In talking to these CEOs, and hearing about their problems, I started to wonder why they weren't hiring older adults. After all, if creativity and intelligence (the roots of innovation) and empathy and wisdom (the roots of collaboration and cooperation) are the superpowers of aging, these CEOs should be going out of their way to hire older candidates.

Two answers rose to the surface: physical fragility and risk aversion.

These CEOs didn't want to hire older adults because their dwindling physical capacities lead to more sick days, less productivity, and too much interruption in work flow. They also felt that older adults were too conservative in their thinking and risk averse in their problem solving, and this combination tended to block innovation rather than speed it along.

But one thing I've learned: Fragility is often a choice. If we train for old age like professional athletes, then we can offset significant physical decline and unlock serious health improvements.

The same is true for risk aversion. Any Gnar Country adventure—where you're going after hard physical goals by consistently pushing on the challenge-skills sweet spot and trying to be creative in your approach—will carve the new neural networks needed to overcome the cautious conservatism that comes with time.

This means by training our bodies and brains to take full advantage of the superpowers that arise with age, we gain access to the very skills that CEOs want most in their companies. In short, if we do "it" right, the over-fifty crowd becomes the dream workforce of the twenty-first century—aka get paid to grow old.

## RULE THIRTEEN: CONSIDER THE GNAR, AKA ACTION SPORTS AS ANTI-AGING MEDICINES

Action sports are fantastic for longevity. All of them are challenging, social, and creative activities that produce feelings of mastery and control. They demand fine-motor performance, fast-twitch muscle response, strength, stamina, balance, agility, flexibility, and a tolerance for risk—that is, the full complement of use it or lose it skills that we need later in life. Also, action sports are packed with flow triggers.

Perhaps most importantly, action sports protect against cognitive decline. The reason is neurogenesis, or the birth of new neurons. If we want to stave off cognitive decline, we need to create new neurons and we need these neurons to form networks—linking new memories to old memories and thus protecting the whole lot against the ravages of time.

In older adults, the hippocampus is where most neurogenesis takes place. The hippocampus is also the part of the brain responsible for map making and long-term memory. It's packed with place cells, grid cells, and other neurons that specialize in recording location.

From an evolutionary perspective, remembering where we were when important stuff happened is critical to survival. Thus, the easiest way to get the hippocampus to birth new neurons is to have emotionally charged experiences in novel and unpredictable environments—which is exactly what action sports provide.

This is why so many mountain towns are Blue Zone longevity hot spots. Summit County, Colorado, is the longest-lived community in America—where the average life span is ten years longer than any other destination. Pitkin County and Eagle County are numbers two and three. They're also in Colorado. All three counties are action-sports meccas, home to Vail, Aspen, Breckenridge, Keystone, Beavercreek, A-Basin, and more.

So the next time the voice in your head says, "You're too old for this shit"—remember, the voice is lying.

The truth? You're too old *not* to do this shit. Action sports are anti-aging medicines and longevity solutions.

Does this mean that you need to learn to park ski? Nope. Try roller-blading, roller-skating, rock climbing, skateboarding, skydiving, scuba diving, snowboarding, spelunking, surfing—and I only bothered to list those action sports that start with the letters *R* and *S*. There a whole alphabet's worth of other possibilities as well.

Does this mean you need to push as hard as I pushed? Or fall on your ass as much as I fell on my ass? Yes and no.

The pushing is mandatory, but the goal is not to push past my limits. It's to push past your own limits. For me, this meant trying to learn to park ski. For you? That's on you to determine.

Meanwhile, those ass falls were on me. But those ass falls were on me for a reason. My dream was to go from absolute beginner to low-level intermediate in a single season. It was such a crazy dream that I didn't bother to put it on my original goal list. Yet once you get to intermediate, you can start to control your progress. A significant amount of fear and randomness get removed from the equation. Sure, you still hit the ground, but it's far less frequent.

As a low-level intermediate park skier, I would suffer fewer injuries and endure less pain. I would also have a near-infinite number of new

ways to get into flow during my very favorite activity in the world. This is how not to lose that brash fire. This is how not to give in to the cozy blanket of middle age. This is how not to go gently into that good night. In other words, this is my version of a retirement plan.

So, you know, Gnar Country. Game on.

SK out.

# Afterword

From a scientific perspective, *Gnar Country* documents a pilot study in peak-performance aging. But it was a rather small study, with just two main participants, Ryan and myself. When I finished writing the book, the only data I had that showed our approach would work for other people was that other people, when they skied with Ryan and myself, would intuitively fall into our game of Chase the Rabbit and often learn freestyle moves along the way. This was certainly tantalizing, but it wasn't evidence.

To remedy this situation, the Flow Research Collective ran a larger study in peak-performance aging in the winter of 2022. Our subjects were a group of seventeen older adults, ages twenty-nine to sixty-eight. In conjunction with two major ski areas, Northstar and Palisades Tahoe, FRC researchers used the principles from this book to teach study subjects how to park ski and park snowboard. With one exception—a man in his late forties recovering from a serious back injury—none of our participants had significant park riding experience. Many were complete novices.

The goal of the study was not to teach our subjects how to throw tricks—though that was one major outcome. Rather, the goal was to teach them to creatively interpret terrain features as a safer and surer path into flow and, by extension, progression. To do that, we broke terrain park riding into eight foundational movements: crouching, jumping, switch riding, slashing, grinding, 180, 360, and a shifty. We spent four days on the mountain. Each day, using a deliberate play/Chase the Rabbit format, subjects learned and practiced two of these movements. Everyone was instructed to gradually build atop previously hardwired motor patterns that could be executed with zero conscious interference and little fear.

To figure out if it worked, we videotaped the training sessions and assessed the results with the same criteria used to judge freestyle competitions by the FIS, the governing body in skiing. We also conducted lengthy preenrollment assessments and interviews, post-study assessments and interviews, and had the subjects take two flow and learning assessments a day, for each of the four days of the study.

To be blunt, the results were freaky. We taught a bunch of old dogs a bunch of new tricks. All our subjects made real progress in most of our judging categories, aka the so-called PAVED criteria: progression, amplitude, variety, execution, and difficulty. We also saw a sizable uptick in flow, which further amplified progression and produced a significant positive shift in attitudes toward later-in-life learning. Afterward, all our participants had reevaluated what they wanted to do with the second half of their lives.

My favorite story about the experiment belongs to my old friend Adam Fisher. Now in his middle fifties, Adam and I met over two decades ago when he was my first editor at *Wired* magazine. We fell out of touch in the intervening years, when Adam went on to a great career as an author, but reconnected around the time I was designing this study. Adam was the perfect subject: an intermediate skier with zero park experience and a considerable amount of fear. Even better, he's one of the most skeptical people I have ever met.

Following class on the second day, after I had departed, Adam, a number of our coaches, and a few students explored the Granite Chief area of Palisades Tahoe. Later that day, I heard a rumor that while back there, Adam jumped off his very first cliff. Cliff jumping is one of the major dividing lines between advanced intermediate and low-level expert. It's a serious milestone for skiers.

I emailed Adam to ask if it was true: "I heard you got busy jumping cliffs in Granite. Bet that was amazing. So much progress and you, of so little faith . . ."

Here's what Adam wrote back:

> You heard true: I think someone even shot a video of that jump! The feeling as I was sailing over the rocks was amazeballs. I didn't do reconnaissance beforehand, just took the coaches' word that the rock (actually many rocks) was there and jumpable "on faith." When I looked down thru my skis as I flew over the rocks, I let out a scream of sheer ecstasy and terror combined. I'm not sure if I have ever felt that way before. It was great. I am actually glad that I didn't look at it before I jumped, because I likely would have chickened out. So yes, much, much progress.
>
> And yes, I was a bit skeptical about flow when we started. Obviously it's a great premise for a book, and obviously flow exists—but the bolder claims that flow can be harnessed on the mountain and then let loose in the workplace seemed to me too good to be true-ish. And yet, riding up Belmont for the nth time, while the rest of Team Geezer was lunching, I had an epiphany of how to structure this manuscript I've been struggling with. And it just popped into my head out of the blue! So skepticism

doesn't even seem to be a factor. Your coaching techniques are working even if I don't "believe" in them!

It's all quite extraordinary, Kotler. You're on to something. . . .

If you want to check out the footage from this experiment or read the postexperiment white paper, you can find both at www.gnar country.com.

If you want to learn the fundamentals of peak-performance aging without having to ski in a terrain park, the Flow Research Collective has you covered. We combined a decade of research and all the major principles in this book to create Enter the Gnar, our core training in peak-performance aging. The course includes a science-based, step-by-step program to help anyone—and especially anyone over forty—set and accomplish high-hard goals. It contains over twenty-five hours of original video content, real-time group coaching, and a battle-tested array of protocols and exercises. If you want to kick ass until you kick the bucket, Enter the Gnar is the place to start. Visit http://www.high flowaging.com to apply. In order to ensure the training is the right fit, you'll be signing up for a quick interview with a member of my team.

While Enter the Gnar tackles peak-performance aging, Zero-to-Dangerous, our flagship flow and peak-performance training, lays the foundation needed to radically elevate your baseline performance and harness flow when you need it most. Tens of thousands of clients across a hundred and thirty countries have used Zero-to-Dangerous to sharpen their focus, amplify their productivity, and reach their goals in record time. You'll come out the other end equipped to accomplish your boldest professional dreams while reclaiming time, space, and freedom in your personal life.

To apply for this course, you can visit https://www.flowresearch collective.com/zero-to-dangerous. In order to ensure the training is the right fit, you'll be signing up for a quick interview with a member of my team.

# Acknowledgments

You cannot go on a Gnar Country adventure without a solid team. Mine was the absolute best. This book would never have been possible without my wife, Joy Nicholson, who handled so much heavy lifting while I was off skiing; Ryan Wickes, who kept me safe, sane, and making steady progress while skiing; and Michael Wharton, who figured out that my peak-performance journal contained the seeds of a book and then helped turn it into this one. Love you all. Also, a shout-out to the Wickes posse—Angela, Kai, Cruz, and Kingston—who backed Ryan and me on the adventure. Wu-Tang is for the children.

On the science front, K. Anders Ericsson got me thinking along these lines. Ongoing conversations with a bunch of neuroscientists, psychologists, and peak-performance experts kept these fires burning. Thanks for the benefit of your brains: Michael Mannino, Mihaly Csikszentmihalyi, Sarah Sarkis, David Eagleman, Adam Gazzaley, Rebecca Rusch, Conor Murphy, Paula Rosales, Laird Hamilton, Gabby Reece, Lynsey Dyer, Andrew Huberman, Chris Malloy, Susi Mai, Rian Doris, Andrew Newberg, Kristen Ulmer, Paul Zak, Clare Sarah, Rich

Diviney, Fred McDaniel, Kele McDaniel, Mark Twight, and Scott Barry Kaufman. Finally, the astounding team at the Flow Research Collective—that's everyone I am lucky enough to work with and everyone we've been lucky enough to train—my deepest gratitude. You are all steazy in the sick gnar!

On the ski hill, much gratitude and many more powder days to Keoki Flagg, Robert Suarez, Sofia Mileti, Tom Day, Eric Arnold, Gordon Fields, Will Kleidon, and Michael Wickes—all of whom made this ride a whole lot more fun. Dirk Collins, Jon Klaczkiewicz, and Jimmy Chin, thanks for that day at Jackson—you know the one. Glen Plake for Hood and for being Glen Plake. John and Dan Egan and Rob and Eric Deslauriers for that Chamonix trip. The OG posse from the Santa Fe Ski Area, especially Dave Stanton, TJ Miller, and Marc Braverman. Everyone involved with Team Geezer, Gnar Country Division. Finally, a loud SICKBIRD to Michael Jaquet, Micah Abrams, the Jerk, Brad Holmes, and the rest of the *Freeze* crew.

On the general inspiration front, a huge debt is owed to everyone at SLVSH.com, especially Joss Christensen and Matt Walker for getting this ball rolling. On the skiing inspiration front, a great many thanks to the entire freestyle community, with an especially deep bow to the folks at Armada, Faction, Moment, ON3P, the Bunch, New Schoolers, Bella Bacon, Alex Hall, Adam Delorme, Eileen Gu, Tom Wallisch, Jake Carney, Henrik Harlaut, Tanner Hall, Alex Hackel, Antti Ollila, Daniel Hanka, Candide Thovex, Phil Casabon, Magnus Graner, Jake Mageau, Jespar Trajer, the crew at the Kimbo Sessions, the killers at the AUDI Nines, the New Canadian Air Force, Wayne Wong, SteepSteep, and everyone else I'm forgetting. Without your shoulders to stand on, we'd all be a lot shorter.

I also want to thank a bunch of other people who helped a bunch along the way. Two old friends—both Paul Bresnick, my agent, and Karen Rinaldi, my publisher at Harper Wave—for believing that *Gnar* was more than a "ski book" and helping make sure it delivered on that promise. Anne Valentino for juggling chain saws. Cynde Christie, Chip Hopper, Vika Viktoria, Sarah Sarkis, and Krista Stryker all read drafts of this book and gave me fantastic notes. The photos were taken by Ryan Wickes and Keoki Flagg—great work, gentlemen, and many thanks. Also, Chris McCann helped tune up images for publication. Alex "Shugz" Dorszynski and Ben Arnst were amazing at coaching our study subjects and filming the results. Everyone at Northstar and Palisades Tahoe who believed in the Gnar Country experiment. All the tireless firefighters who struggled (and struggle) against long odds to keep our forests healthy and mountains safe.

Lastly, to Mom and Dad, for all those rides to the ski hill.

# Notes

## Preface

xi According to Wikipedia, "punk rock": https://en.wikipedia.org/wiki/Punk_rock.

xi Punk embraces a DIY ethic: Rebekah Cordova, *DIY Punk as Education* (Charlotte, NC: Information Age Publishing, 2016). Also see: Kristine Villanueva, "Defining the DIY Scene," Medium, September 25, 2017. For an overall look at the early punk scene: Legs McNeil and Gillian McCain, *Please Kill Me* (New York: Grove Press, 1996).

## Introduction

xiii *The Liftie Report*: https://theliftiereport.epicmountainrentals.com/tlr/the-10-steepest-ski-runs-in-california/.

xiv Thacher Stone: Thacher Stone, "Q&A: You've Seen the Shenanigans, Now Meet the Man Behind @JerryOfTheDay," Freeskier.com, April 27, 2015, https://freeskier.com/stories/youve-seen-the-shenanigans-now-meet-the-man-behind-jerryoftheday. Also see: @jerryoftheday.

xv At most ski areas: There is an ongoing disagreement about terrain signs and slope angles. There are also American, European, and Japanese interpretations. See: Signs of the Mountain, "What Do the Symbols on Ski Trail Signs Mean?," Signsofthemountain.com, https://signsofthemountains.com/blogs/news/what-do-the-symbols-on-ski-trail-signs-mean.

xv steep skiing pioneers: While European skiers invented the "hop-pedal turn" to handle the steeps, most Americans learned the "hop-and-drop" technique, which is sometimes called a "hop turn," from the late Doug Coombs. See Jason Blevins, "Daring Fate," *Denver Post,* April 10, 2006.

xvi Ryan Wickes: All of the Ryan Wickes quotes in this book came from author interviews.

xvi When your brain makes a suggestion: For a lengthy discussion about flow and intuition, see: Steven Kotler, *The Rise of Superman* (New York: New Harvest, 2014), 43–58. Also see: L. Järvilehto, "Intuition and Flow," *Flow Experience* (Dordecht: Springer, 2016), 95–104; A. Bolte et al., "Emotion and Intuition," *Psychological Science* 14, no. 5 (2003): 416–21.

xvii The day the season officially died: Jason Blevins, "The Day Skiing Died: Inside the Historic Day Coronavirus Forced Colorado's Ski Industry to Shutter," *Colorado Sun,* April 15, 2020.

xvii SLVSH videos: SLVSH has its own website at www.slvsh.com, but its YouTube channel is where to find all the games: https://www.youtube.com/c/SLVSH/videos.

## Chapter 1

1 According to traditional learning theories: Gene D. Cohen, *The Creative Age* (New York: Avon Books, 2000), 1–7. Also, much of the credit for ageism in traditional learning theories goes to Sigmund Freud who believed that therapy was useless in anyone over the age of fifty. See: David Smollar, "Freud's Ageism Disputes; Therapy Aids Older People," *Los Angeles Times,* January 26, 1986.

1 Recent discoveries in embodied cognition: Markus Kiefer and Natalie Trump, "Embodiment Theory and Education," *Trends in Neuroscience and Education* (2020): 15–20; and Lawrence Shapiro and Steven Stolz, "Embodied Cognition and Its Significance for Education," *Theory and Research in Education* 17, no. 1 (2019): 19–39. For a general overview of embodied cognition, see: Alva Noe, *Action in Perception* (Cambridge: MIT Press, 2006), 1–35; Andy Clark, *Supersizing the Mind* (Oxford: Oxford University Press, 2011); Margaret Wilson, "Six Views of Embodied Cognition," *Psychonomic Bulletin & Review I* 9, no. 4 (2002): 625–36. For a review of the new field of applied embodied cognition, see: Nuwan Leitan and Lucian Chaffey, "Embodied Cognition and Its Applications: A Brief Review," *Sensoria* 10, no. 1 (2014): 3–10. For a look at how exercise helps the aging brain: Yi-Ping Chao et al., "Cognitive Load of Exercise Influences Cognition and Neuroplasticity of Healthy Elderly," *Journal of Medical and Biological Engineering* 40 (2020): 391–99, and Lavinia Teixeira-Machado et al., "Dance for Neuroplasticity," *Neuroscience & Biobehavioral Reviews* 96 (January 2019): 232–40. Also, the roots of the embodied cognitive approach to learning are often traced to W. Timothy Gallwey, *The Inner Game of Tennis* (New York: Random House, 1974). Finally, for the benefits of both mental and physical training for the brain, see: Clemence Joubert and Hanna Chainay, "Aging Brain: The Effect of Combined Cognitive and Physical Training on Cognition as Compared to Cognitive and Physical Training Alone—A Systematic Review," *Clinical Interventions in Aging* 13 (2018): 1267–301.

1 flow science: For a detailed investigation of all the flow science covered in this book, see: Steven Kotler, *The Art of Impossible* (New York: Harper Wave, 2021), 211–68. For flow as neuroprotective against cognitive decline and ability to accelerate improvement after injury or in the face of illness, see: Thomas Sather et al., "Optimizing the Experience of Flow for Adults with Aphasia," *Topics in Language Disorders* 37, no. 1 (January/March 2017): 25–37; Ji-Hoon Kim, "Influence of Upper Extremity Function, Activities of Daily Living, Therapeutic Flow and Quality of Life in Stroke Patients," *Journal of Digital Convergence* 16, no. 12, 417–26; Kazuki Yoshida et al., "Flow Experience Enhances the Effectiveness of Attentional Training: A Randomized Controlled Trial of Patients with Attention Deficits after Traumatic Brain Injury," *NeuroRehabilitation* 43, no. 2 (2018): 183–93.

1 network neuroscience: For an overview of the major network neuroscience covered in this book, including the "superpowers of aging" ideas, see: Gene D. Cohen, *The Mature Mind* (New York: Basic Books, 2005), 1–49. For the early intersection of network neuroscience and embodied cognitions, there is no better book than J. A. Scott Kelso's *Dynamic Patterns* (Cambridge: Bradford Books, 1997). For some of the studies that changed our thinking about the neurobiological possibilities for older adults, see: J. A. Anguera et al., "Video Game Training Enhances Cognitive Control in Older Adults," *Nature*, September 4, 2013; Denise Park and Gerard Bischof, "The Aging Mind: Neuroplasticity in Response to Cognitive Training," *Dialogues in Clinical Neuroscience* 15, no. 1 (2013): 109–19; Gwenn Smith, "Aging and Neuroplasticity," *Dialogues in Clinical Neuroscience* 15, no. 1 (2013); J. R. Krebs and S. D. Healy, "A Larger Hippocampus Is Associated with Longer-Lasting Special Memory," *PNAS* 98, no. 12 (2001): 6941–44; S. A. Langenecker and K. A. Nielson, "Frontal Recruitment During Response Inhibition in Older Adults Replicated with MRI," *Neuroimage* 20, no. 2 (2003); S. J. Colcombe et al., "Cardiovascular Fitness, Cortical Plasticity, and Aging," *PNAS* 101, no. 9 (2004): 3316–21; A. Alvarez-Buylla and J. M. Gracia-Verdugo, "Neurogenesisi in Adult Subventricular Zone," *Journal of Neuroscience* 22, no. 3 (2002): 619–23; J. Verghese et al., "Leisure Activities and the Risk of Dementia in the Elderly," *New England Journal of Medicine* 348, no. 25 (2003): 2508–16; and F. Nottebohm, "Why Are Some Neurons Replaced in Adult Brains?," *Journal of Neuroscience* 22, no. 3 (2002): 639–43. Also for an additional superpowers of aging, see: Kaoru Nashiro et al., "Age-Related Difference in Brain Activity During Emotion Processing," *Gerontology* 58, no. 2 (February 2012): 156–63.

2 Michael Wharton: Michael Wharton, author interview, 2019.

4 Blue Zones: Dan Buettner, *The Blue Zones* (Washington: National Geographic Partners, 2008). Also see: Thais Abud et al., "Determinants of Heathy Aging: A Systematic Review of Contemporary Literature," *Aging Clinical and Experimental Research*, February 8, 2022; and Francisco Mora, "Successful Brain Aging," *Dialogues in Clinical Neuroscience*, April 1, 2022, 45–52. For an extra look at the serious benefits of intrinsic motivator in the elderly, see: Michiko Sakaki et al., "Curiosity in Old Age: A Possible Key to

Achieving Adaptive Aging," *Neuroscience & Biobehavioral Reviews* 88 (May 2018): 106–16.

4 longevity science: For a general overview of the field of longevity science, see: Steven Kotler and Peter Diamandis, *The Future Is Faster Than You Think* (New York: Simon & Schuster, 2020), 169–79. For a much more detailed examination, see: David Sinclair, *Lifespan* (New York: Atria Books, 2019).

4 five hours of life expectancy: It was CNN anchor Fareed Zakaria who first tracked down the fact that we gain five hours of life expectancy a day. He first mentioned it to me at an event we cohosted in 2015, but he also said it during a speech for the Harvard Alumni Association, see: "Text of Zakaria's Commencement Address," *Harvard Gazette*, May 24, 2012.

5 physical skills begin to decline: Bergita Ganse, Urs Ganse, Julian Dahl, Hans Degens, "Linear Decrease in Athletic Performance During the Human Life Span," *Frontiers in Physiology* (August 21, 2018), Zoran Milanovic et al., "Age-Related Decrease in Physical Activity and Functional Fitness Among Elderly Men and Women," *Clinical Interventions Aging* 8 (May 21, 2018): 549–56.

5 a series of gateways: George Vaillant, *Aging Well* (Boston: Little Brown and Company, 2002), 40–82. Also see: Dilip Jeste and Ellen Lee, "Emerging Empirical Science of Wisdom," *Harvard Revue of Psychiatry* 27, no. 3 (May–June 2019): 127–40.

7 Flow Research Collective: www.flowresearchcollective.com.

8 As body position is tightly linked to embodied cognition: Sandra Blakeslee and Matthew Blakeslee, *The Body Has a Mind of Its Own* (New York: Random House, 2008), 1–15. Also, when it comes to learning new physical skills, the most crucial tie between body position and embodied cognition is memory, see: Lidia Garcia Perez, "Can Body Posture Influence Autobiographical Memory?," *Neuron*, May 12, 2021. Also even though Amy Cuddy's now-famous body posture study has received some criticism, its foundational principles have held up in meta analysis, see: D. R. Carney et al., "Power Posing," *Psychological Science* 21, no. 10 (2010): 1363–68.

9 Adam Delorme: For an introduction to Adam Delorme, see: https://unofficialnetworks.com/2020/04/07/im-thinking-adam-delorme-might-have-the-best-style-in-skiing/.

11 prone to vertigo: For the relationship between Lyme disease, vertigo, and its impact on balance, see: Magdalena Jozefowicz-Korczynska et al., "Vertigo and Severe Balance Instability as Symptoms of Lyme Disease—Literature Review and Case Report," *Frontiers in Neurology* (November 12, 2019).

12 Klingon: There are a number of different Klingon translators around; see: https://www.translator.eu/english/klingon/translation/.

13 older athletes take a lot longer to heal: For a good review of recent findings about recovery times in older athletes, see: MastersAthlete.com.au, "We've Proved It—Older Athletes Do Take Longer to Recover," March 8, 2017. For a look at more current ideas on recovery, essentially the "use it or lose it" notions outlined in this book, see: James Fell and Dafydd Williams, "The Effect of Aging on Skeletal-Muscle Recovery from Exercise: Possible Implications

for Aging Athletes," *Journal of Aging and Physical Activity* 1 (January 16, 2006): 97–115.

13 fine-motor performance: Natasa Miljkovic, Jae-Yong Lim, Iva Miljkovic, Walter Frontera, "Aging of Skeletal Muscle Fibers," *Annals of Rehabilitative Medicine* 39, no. 2 (April 24, 2015): 155–62. For a look at recent research that overturns these ideas, see: Rachael Seidler, "Older Adults Can Learn New Motor Skills," *Behavioral Brain Research* 183, no. 1 (October 1, 2007): 118–22; Claudia Voelcker-Rehage, "Motor-Skill Learning in Older Adults—A Review of Studies on Age-Related Differences," *European Review of Aging and Physical Activity* 5 (January 24, 2008): 5–16. Also see: Sandra Hunter, Hugo Pereira, Kevin Keenan, "The Aging Neuromuscular System and Motor Performance," *Journal of Applied Physiology* 121, no. 4 (October 1, 2016): 982–95; Jonas Leversen, Monika Haga, Hermundur Sigmundsson, "From Children to Adults: Motor Performance Across the Life-Span," *PLoS One* 7, no. 6 (2012). Also, for really early childhood motor skills development, see: Mijna Hadders-Algra, "Early Human Motor Development: From Variation to the Ability to Vary and Adapt," *Neuroscience & Biobehavioral Reviews* 90 (July 2018): 411–27.

19 a use it or lose it situation: Bryant Stamford, "Use Your Muscles or You'll Lose Them," *Courier Journal*, April 3, 2014. Also see: Bergita Ganse, Urs Ganse, Julian Dahl, Hans Degens, "Linear Decrease in Athletic Performance During the Human Life Span," *Frontiers in Physiology* (August 21, 2018); and Pantelis Theodoros Nikolaidis and Beat Knechtle, "Age of Peak Performance in 50-KM Ultramarathoners—Is It Older than in Marathoners?," *Open Access Journal of Sports Medicine* 9 (2018): 37–45.

19 physiologist John Faulkner: John Faulkner, Carol Davis, Christopher Mendias, Susan Brooks, "The Aging of Elite Male Athletes: Age-Related Changes in Performance and Skeletal Muscle Structure and Function," *Clinical Journal of Sports Medicine* 18, no. 6 (November 2008): 501–7.

20 her name was Lola: This is a reference to the song "Copacabana," by Barry Manilow, see: https://en.wikipedia.org/wiki/Copacabana_(song).

## Chapter 2

25 mind wandering as a measure of fitness: Mind wandering is a property of the default-mode network. There are not a lot of studies about the link between DFN deactivation and exercise, but here's one of the more profound ones. Rakib Rayhan et. al., "Exercise Challenge Alters Default Mode Network Dynamics in Gulf War Illness," *BMC Neuroscience* 20, no. 7 (February 21, 2019).

26 flow is the fastest path: For an examination of flow and learning, see: Steven Kotler, *The Art of Impossible* (New York: Harper Wave, 2021). Also: Mihaly Csikszenmihalyi, *Applications of Flow in Human Development and Education* (Dordrecht: Springer, 2014). Also see: Severine Erhel and Eric Jamet, "Improving Instructions in Educational Computer Games: Exploring the Relationship Between Goal Specificity, Flow Experience and Learning Outcomes," *Computers in Human Behavior* 91 (February 2019): 106–14; Lindsay Borovay et al.,

"Flow, Achievement Level, and Inquiry-Based Learning," *Journal of Advanced Academics* 30, no. 1 (2019): 74–106; Sabine Schweder and Diana Raufelder, "Interest, Flow and Learning Strategies: How the Learning Context Affects the Moderating Function of Flow," *Journal of Educational Research* 114, no. 2 (2001): 196–209; Manuel Nianaus et al., "Acceptance of Game-Based Learning and Intrinsic Motivation as Predictors for Learning Success and Flow Experience," *International Journal of Serious Games* 4, no. 30 (September 2017). For a larger cultural perspective: Alanda Maria Ferro Pereira, "Flow Theory and Learning in the Brazilian Context," *Educacao e Pesquisa* 48 (2022).

28 Aduro Sport: https://www.adurosport.com.

29 Lynsey Dyer: Lynsey Dyer, author interview, June 2020. Also see: https://www.lynseydyer.com.

29 women are often better: Aida Grabauskaite et al., "Interoception and Gender," *Consciousness and Cognition* 48 (February 2017): 129–37.

30 Dopamine enhances: For a full breakdown of dopamine and flow triggers, see Steven Kotler, *The Art of Impossible* (New York: Harper Wave, 2021), 233–56. Also, for dopamine and risk, see: Richard Celsi et al., "An Exploration of High-Risk Leisure Consumption Through Skydiving," *Journal of Consumer Research* 20, no. 1 (2013): 1–23. For novelty and uncertainty: C. Teng, "Who Are Likely to Experience Flow?," *Personality and Individual Differences* 50, no. 6 (2011): 863–68.

30 safe, incremental progress: Meysam Beik and Davoud Fazeli, "The Effect of Learner-Adapted Practice Schedule and Task Similarity on Motivation and Motor Learning in Older Adults," *Psychology of Sport and Exercise* 54 (May 2021); Laura Milena Rueda-Delgado et al., "Age-Related Differences in Neural Spectral Power during Motor Learning," *Neurobiology of Aging* 77 (May 2019): 44–57. One interesting detail here, while older and younger brains can learn at the same rate, skill consolidation takes longer in older people, which is another reason older adults need to go slow to go fast. See: K. M. M. Berghuis et al., "Age-Related Changes in Brain Deactivation but Not Activation for Motor Learning," *NeuroImage* 186, no. 1 (February 1, 2019): 358–68. For a look at the idea that older adults have greater unconscious fear loads, see: Lewina Lee and Bob Knight, "Attentional Bias for Threat in Older Adults," *Psychology of Aging* 24, no. 3 (September 2009): 741–47.

31 Unfortunately, these chemicals: For a look at how anxiety can block flow: Amy Rakei et al., "Flow in Contemporary Musicians," *PLoS One* (March 25, 2022). Also see: Jacqueline Rano, Cecilia Friden, Frida Eek, "Effects of Acute Psychological Stress on Athletic Performance in Elite Male Swimmers," *Journal of Sports Medicine and Physical Fitness* 59, no. 6 (June 2019); Susanne Vogel and Lars Schwabe, "Learning and Memory Under Stress," *NPJ Science of Learning* 1 (June 29, 2006). For the relationship between stress and flow, see the literature on the challenge-skills balance, which is often described as flow's most important trigger. Steven Kotler, *The Art of Impossible* (New York, Harper Wave, 2021). Also see: Steven Kotler, *The Rise of Superman* (New York: New Harvest, 2016), 75–90; and Mihaly Csikszentmihalyi, *Flow*

*and the Foundations of Positive Psychology* (Dordrecht: Springer, 2014), 232.

31 Henrick Harlaut: @hharlaut; also see: http://www.xgames.com/athletes/3015422/henrik-harlaut.

32 ski goals could ruin my season: Joannis Zalachoras et al., "Opposite Effects of Stress on Effortful Motivation in High and Low Anxiety Are Mediated by CRHRI in the VTA," *Science Advances* 8, no. 12 (March 23, 2022).

33 Edwin Locke and Gary Lathan: Edwin Locke and Gary Lathan, *Goal Setting* (New Jersey: Prentice Hall, 1984), 10–26. Also, Steven Kotler, *The Art of Impossible* (New York: Harper Wave, 2021), 55–64.

33 brain's primary filters on reality: Maurizio Corbetta, "Control of Goal-Directed and Stimulus-Driven Attention in the Brain," *Nature Neuroscience News* 3 (March 1, 2002): 201–15. Also see: Yuhong Jiang, "Habitual Versus Goal-Driven Attention," *Cortex* 102 (May 2018): 107–20. For fear: Eleonora Vagnoni et al., "Threat Modulates Perception of Looming Visual Stimuli," *Current Biology* 22, no. 19 (October 9, 2012): 826–27; and J. Zaman et al., "Probing the Role of Perception in Fear Generalization," *Scientific Reports,* July 11, 2019.

35 trick list: For a breakdown of freeskiing and freeskiing tricks, check out the "Introduction to Freestyle" video at www.gnarcountry.com.

35 slopestyle: Slopestyle, according to the Olympics: https://olympics.com/en/news/all-you-need-to-know-about-freestyle-skiing.

39 50–50: For a how-to description of the 50–50 and all other tricks in this book, see: https://skiaddiction.com.

42 *Freeze* magazine: For a look at the now defunct *Freeze* magazine, see its Facebook page: https://www.facebook.com/freezemagazine/.

43 punk rock zines: https://en.wikipedia.org/wiki/Punk_zine.

43 mainstream populace hated: For the dislike of punk by the mainstream music industry, also see: Ira Robbins, "Left of the Dial," *Redbull Music Academy Daily,* April 11, 2015. For the transition from punk to new wave, see: https://indie-mag.com/2019/05/new-wave-history/.

43 rebrand the softer: Theodore Cateforis, *Are We Not New Wave* (Ann Arbor: University of Michigan Press, 2011).

44 "Squallywood": Robb Gaffney, *Squallywood* (Sacramento: West Bridge Publishing, 2003).

44 KT-22: https://theliftiereport.epicmountainrentals.com/tlr/the-10-steepest-ski-runs-in-california/. Also see: Stu Campbell, "The Devil's Half-Dozen," *Ski Resort Life,* February 2, 2004.

48 my unnatural direction: Why do we have an unnatural direction? Because humans are side dominant. See: "Science Buddies, Side-Dominant Science," *Scientific American,* February 7, 2013.

49 *Blizzard of Ahhhs*: Greg Stump, *Blizzard of Ahhhs,* Greg Stump Productions, 1988. For a great Q&A with all the core cast members, see: https://www.youtube.com/watch?v=c2rInB8ydO4.

50 Plake wasn't a jock: Rob Hodgetts, "Blizzard of Ahhhs: Punk Antiheroes Launched Skiing's Extreme Generation," *CNN,* December 17, 2018.

51 We were hunting dopamine: For the best research on dopamine and exploration, see: Jaak Panksepp, *Affective Neuroscience* (Oxford: Oxford University Press, 1998), 144–63. Also, V. D. Costa et al., "Dopamine Modulates Novelty-Seeking Behavior During Decision Making," *Behavioral Neuroscience* 128, no. 5 (2014): 556–66.

52 dirty shame: While I don't cover it at length in this book, there appears to be a direct impact of shame on athletic performance; see: Simon Rice et al., "Athlete Experiences of Guilt and Shame," *Frontiers in Psychology*, April 29, 2021.

53 James Jerome Gibson: James Gibson, *The Ecological Approach to Visual Perception* (New Jersey: Lawrence Erlbaum Associates Publishers), 133–35.

## Chapter 3

60 the beginning of an unhealthy obsession: Odin Hjemdal et al., "The Relationship Between Resilience and Levels of Anxiety, Depression and Obsessive-Compulsive Disorder in Adolescents," *Clinical Psychology and Psychotherapy* 18, no. 4 (July–August 2011): 314–21.

60 Buck Brown: https://olympicbootworks.com.

61 retail therapy: Nora Schultz, "Thoughts of Mortality Fuel the Desire for Retail Therapy," *New Scientist* 198, no. 2658 (May 2008): 12.

63 What did Nietzsche say?: Friedrich Nietzsche, *Beyond Good and Evil* (New York: Vintage, 1989), 89.

65 the size of that challenge-skills sweet spot: Sefan Engeser and Falko Rheinberg, "Flow, Performance and Moderators of Challenge-Skill Balance," *Motivation and Emotion* 10, no. 1007 (2008). Also see: William Fernando Garcia, et al., "Dispositional Flow and Performance in Brazilian Triathletes," *Frontiers in Psychology*, September 20, 2019. Finally, further explanation for the shrinkage of the challenge-skills balance can be found in the theory of cognitive reserve; see: Jason Steffener and Yaakov Stern, "Exploring the Neural Basis of Cognitive Reserve in Aging," *Molecular Basis of Disease* 1822, no. 3 (March 2012): 467–73.

65 minimize subconscious fears: Scott Stossel, "The Relationship Between Performance and Anxiety," *Harvard Business Review*, January 6, 2014; and Clive Fullagar, Patrick Knight, Heather Sovern, "Challenge/Skill Balance, Flow, and Performance Anxiety," *Applied Psychology: An International Review* 62, no. 2 (2013): 236–59.

67 German Volume Training: https://fitnessvolt.com/german-volume-training/.

67 Mark Twight: https://www.marktwight.com.

67 Mike Horn: https://www.mikehorn.com.

## Chapter 4

71 Dolores LaChapelle: Dolores LaChapelle, *Deep Powder Snow* (Durango: Kivaki Press, 1993), 32–33.

72 Michael Jordan: The highest vertical leap is an ongoing discussion. For

Michael Jordan's contribution, see: Scott Fujita, "Michael Jordan Vertical Jump," ScottFujita.com, February 8, 2022.

72 long jump: https://en.wikipedia.org/wiki/Long_jump.

76 terrain is ferocious: Barclay, "The 10 Most Challenging Ski Resorts in the United States," *The Unofficial Network*, August 10, 2015.

77 three levels of goals: For a complete breakdown on goal setting, see: Steven Kotler, *The Art of Impossible* (New York: Harper Wave, 2021), 55–64.

79 LL Cool J: LL Cool J, "The Debt," *NCIS Los Angeles*, 2011.

80 played with mirror neurons: For a look at why mimicry enhances learning: Jaclynn Sullivan, "Learning and Embodied Cognition," *Psychology Learning & Teaching* 17, no. 2 (2018): 128–43; Thea Ionescu and Dermina Vasc, "Embodied Cognition: Challenges for Psychology and Education," *Procedia: Science and Behavioral Sciences* 128 (2014): 275–80. For mirror neurons and predictive coding, see: James Kilner, Karl Friston, Chris Frith, "Predictive Coding," *Cognitive Processing* 8 (2007): 159–66; and Angel Lago-Rodriguez et al., "The Role of Mirror Neurons in Observational Motor Learning," *European Journal of Human Movement* 32 (2014): 82–103.

81 Hesitation is death: Arne Nieuwenhuys and Raiul Oudejans, "Anxiety and Perceptual Motor Performance," *Psychological Research* 76, no. 6 (2012): 747–59; Simon Van Gaal et al., "Unconscious Activation of the Prefrontal No-Go Network," *Journal of Neuroscience* 30, no. 11 (March 17, 2010): 4143–50.

83 know who you are and how you like to learn: Chris Jackson and Michele Lawty-Jones, "Explaining the Overlap Between Personality and Learning Style," *Personality and Individual Differences* 20, no. 3 (1996): 293–300. Also, Jessica Heinström, "Personality Effects on Learning," in N. M. Seel, ed., *Encyclopedia of the Sciences of Learning* (Boston: Springer, 2012). Also see: Sabine Schweder and Diana Raufelder, "Interest, Flow and Learning Strategies: How the Learning Context Affects the Moderating Function of Flow," *Journal of Education Research* 114, no. 2 (March 23, 2020): 196–209.

86 in the language of motor learning: Janelle Weaver, "Motor Learning Unfolds Over Different Timescales in Distinct Neural Systems," *PLoS Biology* 13, no. 12 (December 2015). Also, Paul Fitts and Michael Posner, *Human Performance* (Oxford: Brooks/Cole, 1967).

87 repetition suppression: Lisa Mayehauser et al., "Neural Repetition Suppression," *Frontiers in Human Psychology*, April 17, 2014.

87 the brain reduces the wow: David Eagleman et al., "Predictability Engenders More Efficient Neural Responses," *Nature*, February 3, 2009.

90 that roar worked: Pawel Fedurek et al., "The Relationship Between Testosterone and Long-Distance Calling in Wild Chimpanzees," *Behavioral Ecology and Sociobiology* 70, no. 5 (May 2016); and Sari van Anders, Jeffrey Steiger, Katherine Goldey, "Effects of Gendered Behavior on Testosterone in Men and Women," *PNAS* 112, no. 45 (October 26, 2015): 13805–10.

92 a "schema": Schema theory has a colorful history inside the field of motor learning, see: Richard Schmidt, "Motor Schema Theory After 27 Years," *Research Quarterly for Exercise and Sport* 74, no. 4 (2003): 366–75.

## Chapter 5

95 one of the bigger terrain parks in Tahoe: Snowledge Team, "Northstar Top to Bottom: Devan Peeters Greases 25 Features in 3 Minutes," Snowledge .com, May 26, 2017. See: https://www.snowledge.co/blog/northstar-terrain -parks/.

98 "the Fingers": For a look at the Fingers and many other lines mentioned in this book: https://snowbrains.com/scratching-the-technical-itch-kirkwood/.

102 Little wins produce dopamine: Sefano di Domenico and Richard Ryan, "The Emerging Neuroscience of Intrinsic Motivation," *Frontiers in Human Neuroscience*, March 24, 2017. Also see: Elliot Berkman, "The Neuroscience of Goals and Behavior Change," *Consult Psychology Journal* 70, no. 1 (March 2018): 28–44.

103 the neurochemicals that underpin: Sutton Harber, "Endorphins and Exercise," *Sports Medicine* 1, no. 2 (March 1984): 154–71; and Sonja Vuckovic et al., "Cannabinoids and Pain," *Frontiers in Pharmacology* 9 (2018).

103 potent pain relievers: Steven Kotler, *The Rise of Superman* (New York: New Harvest, 2014), 66–67. Also: Johannes Fuss et al., "A Runner's High Depends on Cannabinoid Receptors in Mice," *PNAS* 112, no. 42 (October 5, 2015); Arne Dietrich and William McDaniel, "Endocannabinoids and Exercise," *British Journal of Sports Medicine* 38 (2004): 536–41.

106 "deliberate practice": K. Anders Ericsson, "Deliberate Practice and the Acquisition and Maintenance of Expert Performance in Medicine and Related Domains," *Academic Medicine* 79, no. 10 (October 2004): 70–81.

106 Well, Anders: K. Anders Ericsson, "The Influence of Experience and Deliberate Practice on the Development of Superior Expert Performance," *The Cambridge Handbook of Expertise and Expert Performance* (Cambridge: Cambridge University Press, 2006) 685–705.

109 The day started with *The Art of Impossible*: https://www.theartofimpossible .com.

113 flow triggers: For a technical examination of the triggers for group flow, see: Jef J. J. van den Hout and Orin Davis, *Team Flow* (New York: Springer, 2019). For a more general overview, see Keith Sawyer, *Group Genius* (New York: Basic Books, 2007); and for all the triggers together, see Steven Kotler, *The Art of Impossible* (New York: Harper Wave, 2021), 233–56.

116 And what does group flow look like: For a group flow overview, see: Keith Sawyer, *Group Genius* (New York; Basic Books, 2007), and Steven Kotler, *The Rise of Superman* (New York: New Harvest, 2014), 127–49.

119 There was a U2: https://www.u2.com/lyrics/111.

121 Nailing our performance: For a really nice article about the role of dopamine and performance in front of an audience, see: Peachy Essay, "Theater Performance and Media Result of Neurochemistry Changes in the Brain," Peachyessay.com, January 10, 2021. For a more technical description: Vikram Chib, Roy Adachi, John O'Doherty, "Neural Substrates of Social Fascination Effects on Incentive-Based Performance," *Social Cognitive and Affective Neuroscience* 13, no. 4 (April 2018): 391–403.

121 show-offs like to show off: Marie Banich and Stan Floresco, "Reward Systems, Cognition, and Emotion: Introduction to the Special Issue," *Cognitive, Affective, & Behavioral Neuroscience* (May 23, 2019): 409–14.

122 nonverbal lexicon: Walter Cannon, *Bodily Changes in Pain, Hunger, Fear and Rage* (New York: Appleton, Century, Crofts, 1929).

123 Exploratory mode: Colin DeYoung, "The Neuromodulator of Exploration: A Unifying Theory of the Role of Dopamine in Personality," *Frontiers of Human Neuroscience*, November 14, 2013.

126 Andrew Huberman: Andrew Huberman discusses the relationship between vision and fear all over the web. Two personal favorites are his conversation with *Scientific American* (Jessica Wapner, "Vision and Breathing May Be the Secret to Surviving 2020," *Scientific American*, November 16, 2020) and in conversation with former Navy SEAL commander Rich Diviney and the Flow Research Collective, https://www.flowresearchcollective.com/radio/6.

127 Alia Crum: Parul Somani, "Mindsets: Q&A with Dr. Alia Crum, Stanford Psychology," ParulSomani.com, December 31, 2019.

127 Research dating back to the 1970s: Most of the early research on mindset was conducted by Harvard psychologist Ellen Langer. For a great roundup of this work, including her most famous "counterclockwise study," see: Ellen Langer, *Counterclockwise: Mindful Health and the Power of Possibility* (New York: Random House, 2009).

127 Ohio Longitudinal Study: Robert Atchley, "Ohio Longitudinal Study on Aging and Retirement, 1975–1995," https://doi.org/10.7910/DVN/XL2ZTO, Harvard Dataverse, V1.

128 addictiveness of progress: For a broad introduction to the relationship between dopamine and compulsive behavior, including success, see: David Linden, *The Compass of Pleasure* (New York: Viking, 2011).

## Chapter 6

131 five major intrinsic motivators: Steven Kotler, *The Art of Impossible* (New York: Harper Wave, 2021), 17–97.

131 three tiers of goals: Ibid.

137 Marijuana decreases inflammation: For a general look at the endocannabinoid system and stress, see: Ryan Wyrofsky et al., "Endocannabinoids, Stress Signaling, and the Locus Coeruleus-Norepinephrine System," *Neurobiology of Stress*, November 11, 2019. For an overview of recent research: Emily Earlenbaugh, "New Research Reveals Why Cannabis Helps PTSD Sufferers," *Forbes*, September 17, 2020. For a look at the science: Marcel Bonn-Miller et al., "The Short-Term Impact of Three Smoked Cannabis Preparations Versus Placebo on PTSD Symptoms," *PLoS One*, March 17, 2021; Christine Rabinak et al., "Cannabinoid Facilitation of Fear Extinction Memory Recall in Humans," *Neuropharmacology* 64 (2013): 396–402; Leah Mayo et al., "Targeting the Endocannabinoid System in the Treatment of Post-Traumatic Stress Disorder," *Biological Psychiatry* 91, no. 3 (February 2022): 262–72.

139 Sense of meaning: Flow's impact on meaning is well documented. I cover it

thoroughly in *The Art of Impossible*, but if you want a detailed look at flow on human development, see: Mihaly Csikszentmihalyi, *Applications of Flow in Human Development and Education* (New York: Springer, 2014), 14–18, 31, 83.

140 embodied cognition: Jennifer Fugate et al., "The Role of Embodied Cognition in Transforming Learning," *International Journal of School and Education Psychology*, 2018.

140 anger did its job: Grace Giles et al., "When Anger Motivates," *Frontiers in Psychology*, August 5, 2020.

143 Grit is a limited resource: Roy Baumeister, *Willpower* (New York: Penguin, 2012).

## Chapter 7

145 *deliberate play*: The literature on learning and play is considerable. For a good review: Claire Liu et al., "Neuroscience and Learning Through Play: A Review of the Evidence," LEGO Foundation, November, 2017. The LEGO Foundation has published an excellent summary of this work and more, *Learning Through Play: A Review of the Evidence,* which is available here: https://cms.learningthroughplay.com/media/0vvjvscx/learningthroughplay_areview_summary.pdf. For a look at the neurobiology of play: Stephen Siviy and Jaak Panksepp, "In Search of the Neurobiological Substrates for Social Playfulness in Mammalian Brains," *Neuroscience & Biobehavioral Reviews* 35, no. 9 (October 2011): 1821–30. Also see: Rene Proyer, "The Well-Being of Playful Adults," *European Journal of Humour Research* 1, no. 1 (2013); and Cale Magnuson and Lyn Barnett, "The Playful Advantage," *Leisure Sciences* 35, no. 2 (2013): 129–44. Lastly, Stuart Brown's *Play* (New York: Avery, 2008) remains a great book on the entire subject.

146 the amygdala: For a look at how mindset impacts motor learning in adults: Gabriele Wulf et al., "Altering Mindset Can Enhance Motor Learning in Older Adults," *Psychology and Learning* 27, no. 1 (2012): 14–21.

147 *Crouching Tiger:* https://www.imdb.com/title/tt0190332/.

147 Keoki Flagg: https://www.gallerykeoki.com.

148 Flow is a four-stage cycle: Jeffery Dusek and Herbert Benson, "Mind-Body Medicine," *Minnesota Medicine* 92, no. 5 (May 2009): 47–50. Jeffery Dusek et al., "Association Between Oxygen Consumption and Nitric Oxide Production During the Relaxation Response," *Medical Science Monitoring* 12, no. 1 (January 2006). Also see: Steven Kotler, *The Art of Impossible* (New York: Harper Wave, 2021), 257–68.

151 a midair suggestion: The best examination of how the brain performs during moments of improvisation can be found in John Kounios and Mark Beeman, *The Eureka Factor* (London: Windmill Books, 2015). How this work directly relates to flow is covered in *The Art of Impossible*, 175–90.

160 "Yes, and": Psychologist Keith Sawyer did all of the early research on "yes, and" as a flow trigger. Keith Sawyer, *Group Genius* (New York: Basic Books, 2007). For a detailed look at how group flow impacts athletic performance,

see: Steven Kotler, *The Rise of Superman* (New York: New Harvest, 2014), 129–48.

162 Tom Wallisch: Megan Michelson, "Tom Wallisch," *The Ski Journal*, see: https://www.theskijournal.com/exclusive/tom-wallisch/.

## Chapter 8

165 Fred McDaniel: https://www.humanperformancecenter.com. Also, Fred McDaniel and his wife, Kele, coauthored the definitive guide to flexibility for cyclists. In my experience, it also works great for skiing and all other action sports. You can find it on their website. Also see: Nick Heil, "Reprogramming Your Fitness Brain," *Outside*, October 5, 2009.

167 Tom Day: SnowBrains did a great podcast episode with Tom Day, not long after Tom won his Emmy. See https://snowbrains.sounder.fm/episode/tom-day. Also see IMDB for complete filmography, https://www.imdb.com/name/nm2497471/.

170 Mammoth sits at the epicenter: https://www.mammothmountain.com/unbound-terrain-parks.

171 full menu of affordances: James Gibson, *The Ecological Approach to Visual Perception* (New Jersey: Lawrence Erlbaum Associates, 1986), 147. It's also worth noting that Gibson published his original theory of affordances in 1979, and it has changed over time. For a more recent review, see: Harold Jenkins, "Gibson's Affordances," *Journal of Scientific Psychology*, December 2008. For a direct look at how affordances impact athletics, see: Brett Fajen, Michael Riley, Michael Turvey, "Information, Affordances, and the Control of Action in Sport," *International Journal of Sport Psychology* 40, no. 1 (November 2008). For a look at how affordances impact learning in action and adventure sports, see: Ludovic Seifert, Guillaume Hacques, John Komar, "The Ecological Dynamics Framework: An Innovative Approach to Performance in Extreme Environments," *International Journal of Environmental Research and Public Health* 19, no. 5 (February 2022).

171 piques my curiosity: For curiosity as a motivator and in relation to flow, see: Stefano Di Domenico and Richard Ryan, "The Emerging Neuroscience of Intrinsic Motivation," *Frontiers of Human Neuroscience*, March 24, 2017.

172 Joss Christensen: @joss. Also see: https://www.teamusa.org/Athletes/CH/Joss-Christensen.

175 My chances of dying seemed minuscule: While the literature around pattern recognition and threat detection is fairly thick, it becomes thin when you're trying to understand how the brain downgrades a threat. Here, when my brain noticed the berms were huge and my chances of dying were small, this is a moment of pleasant surprise, what is considered a better than expected outcome in the literature of predictive coding. For a look at how this works, see: Greg Berns et al., "Predictability Modulates Human Brain Response to Reward," *Journal of Neuroscience* 21, no. 8 (April 15, 2001): 2793–98.

175 A bunch of sensations: Technically, when performance anxiety mounts over time it can produce "stress-response hyperstimulation," which directly

impacts muscle tightness, soreness, and fatigue. For a general overview, see https://www.anxietycentre.com/anxiety-disorders/symptoms/hyperstimulation/.

176 Will Kleidon: Will Kleidon is the CEO of Ojai Energetics. He is an expert in cannabis and the endocannabinoid system. On the relationship between flow, cannabis, and creativity, the Flow Research Collective Radio hosted a great discussion between myself, futurist Jason Silva, and Will Kleidon. See https://flowresearchcollectiveradio.podbean.com/e/16/.

177 "flooding": For a review of systematic desensitization, which is the technical term for the approach I took to overcoming vertigo at Mammoth, see: Saul McLeod, "Systematic Desensitization as a Counter-Conditioning Process," *Simple Psychology*, www.simplepsychology.org. I also describe this approach in *The Art of Impossible*, 83–89.

178 battled anxiety with curiosity: When you use curiosity and excitement to counteract anxiety, this is known as either "reframing" or "anxious reappraisal." For a great overview of recent research, see: Olga Khazan, "Can Three Words Turn Anxiety into Success," *The Atlantic*, March 23, 2016.

180 courage is essentially dopamine: Andrew Huberman, author interview, 2020. The discovery of the relationship between courage and dopamine came from Dr. Andrew Huberman's lab at Stanford. Also see: Bruce Goldman, "Scientists Find Fear, Courage Switches in the Brain," *Stanford Medicine*, May 2, 2018.

182 $VO_2$ max: Brad Stulberg, "Endurance Guru Joe Friel Says You Can Still Be Fast After 50," *Outside*, March 4, 2015.

182 another use it or lose it skill: Vivian Giang, "You Can Teach Yourself to Be a Risk-Taker," BBC, June 6, 2017.

183 openness to experience: Jill Suttie, "What Neuroscience Can Teach Us About Aging Better," *Greater Good*, January 20, 2020. Also see: Tess Gregory et al., "Openness to Experience, Intelligence, and Successful Aging," *Personality and Individual Differences* 48, no. 8 (June 2010): 859–99.

## Chapter 9

188 another use it or lose it situation: Gary Hunter, John McCarthy, Marcas Bamman, "Effects of Resistance Training on Older Adults," *Sports Medicine* 34, no. 5 (2004): 329–48.

189 "jumper's knee": For an overview of jumper's knee in freestyle skiing see, "Fucking Patellar Tendonitis," at New Schoolers, https://www.newschoolers.com/forum/thread/755856/Fucking-Patellar-Tendonitis.

191 keep on learning later in life: For an overview of the impact of late-in-life learning on mental and physical health, see Gene D. Cohen, *The Mature Mind* (New York: Basic Books, 2005).

192 Old is a mindset: Ellen Langer, *Counterclockwise: Mindful Health and the Power of Possibility* (New York: Random House, 2009).

192 stuff that works like ice: Yes, there is now a giant controversy surrounding ice. There are old schoolers who believe in its healing powers, and new schoolers

who believe that reducing inflammation after injury (which is what ice does) slows and sometimes blocks healing. You can find a ton of information online by searching for "the ice debate." I tend to side with the pro-ice camp, because the anti-ice evidence is still less than convincing.

193 The first time I experimented with regenerative medicine: If you're curious about the early history of regenerative medicine, I reported on the subject for the *LA Weekly*. The article is reprinted in Steven Kotler, *Tomorrowland* (New York: New Harvest, 2015), 183–200.

193 platelet-rich plasma: PRP therapy is also controversial. For a favorable look at recent evidence, see: Peter Everts et al., "Platelet-Rich Plasma: New Performance Understandings and Therapeutic Considerations in 2020," *International Journal of Molecular Science* 21, no. 20 (October 2020). For a thoroughly damning review of the evidence, see: Kade Paterson, "Cutting Through the Hype on Platelet-Rich Plasma," *Pursuit*, University of Melbourne, January 13, 2022. Personally, I believe the real issue isn't with the therapy itself, but with the skill of the doctor making the injections. As far as I can tell, reading an ultrasound and using that image to guide an injection is more of an art than a science. Additionally, I found PRP useful in treating shoulder injuries, semiuseful in treating back injuries, and not useful in treating ankle injuries. I'm not alone. In talking to other athletes about PRP, you frequently hear stories about similarly mixed results.

194 Dr. Matt Cook: Dr. Matt Cook has a podcast, and it's a great place to start if you're trying to get a sense of his approach to regenerative medicine: https://bioresetpodcast.com.

194 Exosomes: Like everything else in regenerative medicine, exosomes themselves are controversial, but there is a mountain of evidence for the level of effectiveness I experienced. See: Donald Phinney and Mark Pittenger, "Concise Review: MSC-Derived Exosomes for Cell-Free Therapy," *Stem Cell Express*, March 7, 2017. Also: Wumei Wei et al., "Mesenchymal Stem Cell-Derived Exosomes," *Frontiers in Pharmacology*, January 25, 2021.

198 Jeremy Jones: @jeremyjones. Also, Jones founded the excellent organization Protect Our Winters to battle climate change, see: https://protectourwinters.org.

203 Wittgenstein: Ludwig Wittgenstein, *Tractatus Logico-Philisophicus* (New York: Harcourt, Brace, 1933), 68.

## Chapter 10

205 I designed my one-inch-at-a-time approach to promote: In my experience, the one-inch-at-a-time protocol that underpins my Gnar Country experiment helps reduce catastrophic injuries, but it can't prevent them (as my T-boning incident definitely illustrated). Also, by catastrophic, I am referring to anything that requires surgery to heal. There is also a much worse category—permanent injury or the kind that surgery still can't heal. All of this is to say, my approach is dangerous and can result in serious injury, even death. Proceed with serious caution.

205 Regenerative medicine seemed ready to handle: The belief that regenerative medicine is ready to handle chronic injuries is mine alone, and entirely based on personal experience. I've had success using it to treat knees (MCL tears and patellar tendinitis), shoulders (rotator cuff tears), and back problems (arthritis). Will you have the same success? Honestly, I have no idea. It's also worth pointing out that regenerative medicine is not yet covered by insurance and the costs are extremely high. The good news here is that the evidence for its success continues to mount and the treatment is becoming more popular—thus costs are starting to come down. Additionally, while I have focused on exosomes, there are less expensive treatments available via peptides. This too is both new and controversial, but again the evidence for peptides' ability to regenerate tissues and organs is mounting. See, Matthew Webber and Samuel Stupp, "Emerging Peptide Nanomedicine to Regenerate Tissues and Organs," *Journal of International Medicine* 267, no. 1 (January 2010): 71–88.

211 the five Blue Zone keys: Dan Buettner, *The Blue Zones* (Washington, DC: National Geographic, 2008), 261–98.

212 In the elderly, leg strength: Anne Newman et al., "Strength, but Not Muscle Mass, Is Associated with Mortality in Health, Aging and Body Composition Study Cohort," *Journals of Gerontology* 61, no. 1 (January 2006): 72–77. Also see: Antonio Garcia-Hermoso et al., "Muscular Strength as a Predictor of All-Cause Mortality in an Apparently Healthy Population," *Archives of Physical and Medical Rehabilitation* 99, no. 10 (October 2018).

212 common killer of older adults: The Endocrine Society, "Broken Bones Among Older People Increase Risk of Death for Up to 10 Years: Femur, Pelvic Fractures Pose Similar Risk as Hip Fractures," *ScienceDaily*, July 19, 2018.

212 researchers at Kings College London: Claire Steves et al., "Kicking Back Cognitive Ageing: Leg Power Predicts Cognitive Ageing After Ten Years in Older Female Twins," *Gerontology* 62, no. 2 (2016): 138–49.

212 inflammation, which is the root of much that we call "aging": Luigi Ferrucci and Elisa Fabbri, "Inflammaging: Chronic Inflammation in Aging, Cardiovascular Disease, and Frailty," *Nature Reviews Cardiology* 15 (2018): 505–22; Hae Young Chung et al., "Redefining Chronic Inflammation in Aging and Age-Related Diseases," *Aging and Disease* 10, no. 2 (April 2019): 367–82; Helen Lavretsky and Paul Newhouse, "Stress, Inflammation and Aging," *American Journal of Geriatric Psychiatry* 20, no. 9 (September 2012): 729–33. Also see: Nicholas Justice, "The Relationship Between Stress and Alzheimer's Disease," *Neurobiology of Stress*, April 21, 2018.

212 Stress weakens motivation: Nick Hollon, Lauren Burgeno, Paul Philips, "Stress Effects on the Neural Substructures of Motivated Behavior," *Nature Neuroscience* 18, no. 10 (October 2015): 1405–12.

212 blocks flow: The relationship between stress and flow is complicated. The most obvious place it shows up in is the tuning of the challenge-skills balance, wherein too much anxiety pushes one out of the balance and makes achieving flow much more difficult. See Steven Kotler, *The Art of Impossible* (New York: Harper Wave, 2021). Also, the research on how stress blocks

creativity is thoroughly explored in Mark Beeman and John Kouinos, *The Eureka Factor* (London: Windmill Books, 2015).

212 Time in nature: The relationship between time in nature and mental health is so well established that there is an entire field of research devoted to its understanding; see: M. G. Berman, A. J. Stier, and G. N. Akcelik, "Environmental Neuroscience," *American Psychologist* 74, no. 9 (2019): 1039–52. Also, there are now meta-analyses that have figured out what "time dose" (aka time spent outdoors) produces the best mental health benefits. For a thorough review, see Genevive Meredith et al., "Minimum Time Dose in Nature to Positively Impact the Mental Health of College-Aged Students, and How to Measure It: A Scoping Review," *Frontiers in Psychology*, January 14, 2020. Finally, for a look at outdoor activity and health in older adults, see: Jacqueline Kerr et al., "The Relationship Between Outdoor Activity and Health in Older Adults Using GPS," *International Journal of Environmental Research and Public Health* 9, no. 12 (2012): 4615–25.

214 use the trampoline park to increase my air sense: Joshua Aman et al., "The Effectiveness of Proprioceptive Training for Improving Motor Function: A Systematic Review," *Frontiers in Human Neuroscience*, January 28, 2015. Also, Fabian Herold et al., "Thinking While Moving or Moving While Thinking—Concepts of Motor-Cognitive Training for Cognitive Performance Enhancement," *Frontiers in Aging and Neuroscience*, August 6, 2018.

214 my attempt to lateralize: For lateralization and skill acquisition, see Steven Kotler, *The Rise of Superman* (New York: New Harvest, 2014), 123–26.

214 "risk tolerance": Thomas Dohmen et al., "Identifying the Effect of Age on Willingness to Take Risks," *CEPR*, January 21, 2018.

215 "match quality": For the best description of match quality, see: David Epstein, *Range* (New York: Riverhead Books, 2019), 128.

215 put down resentments: Javier Lopez et al., "Forgiveness Interventions for Older Adults," *Journal of Clinical Medicine* 10, no. 9 (May 2021). See also N. J. Webster, K. J. Ajrouch, T. C. Antonucci, "Towards Positive Aging: Links Between Forgiveness and Health," *OBM Geriatrics* 4, no. 2 (2020): 118.

215 Fear blocks empathy and decreases creativity: For fear and empathy, see: Andrew Todd et al., "Anxious and Egocentric," *Journal of Experimental Psychology: General* 144, no. 2 (April 2015): 374–91. For the relationship between fear and creativity, see John Koinios and Mark Beeman, *The Eureka Factor* (London: Windmill Books, 2015), 115–19.

215 further lowers our appetite for risk: Anthony Porcelli and Mauricio Delgado, "Acute Stress Modulates Risk Taking in Financial Decision Making," *Psychological Science* 20, no. 3 (March 2009). Also, Mara Mather and Nichole Lighthall, "Both Risk and Reward Are Processed Differently in Decisions Made Under Stress," *Current Directions in Psychological Science* 21, no. 2 (March 26, 2012).

217 Gene Cohen: Gene D. Cohen, *The Creative Age* (New York: Avon Books, 2000); Gene D. Cohen, *The Mature Mind: The Positive Power of the Aging Brain* (New York: Basic Books, 2006); Gene D. Cohen, "The Creativity and Aging Study," George Washington University, 2006.

219 Sarah Sarkis: Author interview with Dr. Sarah Sarkis, June 2021. Also see: https://drsarahsarkis.com. Also, for the Flow Research Collective Radio interview with Dr. Sarkis, see: https://drsarahsarkis.com/2020/12/28/rethinking-cognitive-bias-and-the-unconscious-the-frc-podcast-episode/.

220 Margaret Atwood: I'm paraphrasing here, the actual quote is: "Everyone else my age is an adult, whereas I am merely in disguise." Margaret Atwood, *Cat's Eye* (New York: Anchor, 1998), 15.

221 Laird Hamilton: Author interview with Laird Hamilton, June 2021.

## Appendix: The Rules

228 In high-risk situations, the brain: Rongjun Yu, "Stress Potentiates Decision Biases," *Neurobiology of Stress* 3 (June 2016): 83–95. Also, Sean Wake et al., "The Influence of Fear on Risk-Taking," *Cognition and Emotion* 34, no. 6 (2020).

228 where the brain sets the challenge-skills balance: Susan Jackson, "Factors Influencing the Occurrence of Flow State in Elite Athletes," *Journal of Applied Sport Psychology* 7, no. 2 (1995): 138–66; and Susan Jackson et al., "Psychological Correlates of Flow in Sport," *Journal of Sport and Exercise Psychology* 20, no. 4 (1998): 358–78. Also see: Edward Chavez, "Flow in Sport: a Study of College Athletes," *Imagination, Cognition and Personality* 28, no. 1 (January 21, 2009): 69–91; Stefan Koehn, "Pre-Performance Confidence as a Predictor of Flow State," *Medicine and Science in Tennis* 17, no. 1 (February 2012): 16–21.

230 triggers for group flow: Jef van den Hout and Orin Davis, *Team Flow* (Switzerland: Springer, 2019), 31–49. Also Keith Sawyer, "What Mel Brooks Can Teach Us About 'Group Flow,'" *Greater Good Magazine*, January 24, 2012.

230 maintaining healthy interpersonal relationships: Katherine Harmon, "Social Ties Boost Survival by 50 Percent," *Scientific American*, July 28, 2010. Also see: Yang Claire Yang et al., "Social Relationships and Physiological Determinants of Longevity Across Human Life Span," *PNAS* 113, no. 3 (January 19, 2016): 578–83. For flow and social bonding, see: Jason Keeler et al., "The Neurochemistry and Social Flow of Singing: Bonding and Oxytocin," *Frontiers in Human Neuroscience*, September 25, 2015.

231 the line between "fear" and "too much fear": For the complicated relationship between fear and too much fear, see: Ilse Van Diest, "Interoception, Conditioning and Fear," *Psychophysiology* 56, no. 8 (August 2019). For the relationship between conscious and unconscious fear and interoception: Kai MacDonald, "Interoceptive Cues: When Gut Feelings Point to Anxiety," *Current Psychiatry* 6, no. 11 (November 2007).

231 the dopamine generated by pattern recognition: Sergei Gepshtein et al., "Dopamine Function and the Efficiency of Human Movement," *Journal of Cognitive Neuroscience* 26, no. 3 (March 2014): 645–57. For a look at how dopamine amplifies psychological and physical factors, see Pat Davidson, "Central Fatigue and the Role of Neurotransmitters on Reduced Work Output," Simplifaster.com, https://simplifaster.com/articles/central-fatigue-role-neurotransmitters-reduced-work-output/.

232 "Older persons": Gene D. Cohen, *The Mature Mind* (New York: Basic Books, 2006), 27. Also see: University of Toronto, "Old Brains Can Learn New Tricks," *Science Daily*, October 25, 1999.

232 This health boost comes from: Joanna Hong et al., "The Positive Influence of Sense of Control on Physical, Behavioral, and Psychosocial Health in Older Adults," *Preventative Medicine* 149 (August 2021).

232 sense of control: Ellen Langer and Judith Rodin, "The Effect of Choice and Enhanced Personal Responsibility for the Aged: A Field Experiment in an Institutional Setting," *Journal of Personality and Social Psychology* 34, no. 2 (September 1976): 191–98.

232 Positive emotions increase: Fulvio D'Acquisto, "Affective Immunology," *Dialogues in Clinical Neuroscience* 19, no. 1 (March 19, 2017): 9–19. Also, B. Easterling et al., "Psychosocial Modulation of Cytokine-induced Natural Killer Cell Activity in Older Adults," *Psychosomatic Medicine* 58, no. 3 (May–June 1996): 264–72.

233 how we create *a whole lot more life*: For the details about well-being and overall life satisfaction, see: Mihaly Csikszentmihalyi, *Applications of Flow in Human Development and Education* (Cham, Switzerland: Springer, 2014), 24–32. Additionally, Martin Seligman's *Authentic Happiness* (New York: Free Press, 2002) is another great resource.

234 Flow fights anxiety: The research into the inverse relationship between flow and anxiety traces back to Herbert Benson, *The Breakout Principle* (New York: Scribner, 2003). For a more contemporary update, see: US Pain Foundation, "How Flow State and the Nervous System Interact," USPainFounda tion.org, July 28, 2021. Also, for my own experience with how flow can reset the nervous system and boost the immune system, see: Steven Kotler, *West of Jesus* (New York: Bloomsbury, 2006).

234 positive emotions stimulate the immune system and slow aging: For a general overview of flow's impact on healthy aging, see: Miguel Bautista, "Flow State, Exercise and Healthy Aging," *The Conversation*, April 13, 2022.

235 All of the performance-enhancing neurochemicals . . . tie to emotions: For the relationship between emotions and neurochemistry, see: Jaak Panksepp, *Affective Neuroscience* (Oxford: Oxford University Press, 1998), and Lisa Feldman Barrett, *How Emotions Are Made* (Boston: Mariner Books, 2018).

235 Flow amplifies our ability . . . to remember what we did: Jennifer Schmidt, "Flow in Education," *Education*, 2010. Chris Berka et al., "EEG Correlates of Task Engagement and Mental Workload in Vigilance, Learning, and Memory Tasks," *Aviation, Space, and Environmental Medicine* 78, no. 5 (2007): B231–44. Also Kevin Rathunde and Mihalyi Csikszentmihalyi, "Middle School Students' Motivation and Quality of Experience: A Comparison of Montessori and Traditional School Environments," *American Journal of Education* 111, no. 3 (2007): 341–71. Also S. Craig, A. Graesser, J. Sullins, and B. Gholson, "Affect and Learning: an Exploratory Look into the Role of Affect in Learning with Autotutor," *Journal of Educational Media* 29, no. 3 (October 2004): 241–50. Finally, see: Susan Jackson and Mihaly Csikszentmihalyi,

*Flow in Sport: The Keys to Optimal Experiences and Performances* (Champaign, IL: Human Kinetics, 1999), 65–68.

236 Recover Like a Pro: Peter Reaburn, Matthew Driller, and Christos Argus, "Age-Related Changes in Performance and Recovery Kinetics in Masters Athletes: A Narrative Review," *Journal of Aging and Physical Activity* 24, no. 1 (January 2016). For a general overview of training and recovery tips from older pro athletes, see Eric Benson, "How Pro Athletes Like LeBron James and Tom Brady Are Playing Longer (and Better) Than Ever," Men's Journal.com, https://www.mensjournal.com/sports/how-todays-ageless-pros-are-reaching-athletic-immortality/.

236 train for old age like a professional: James McKendry et al., "Comparable Rates of Integrated Myofibrillar Protein Synthesis Between Endurance-Trained Master Athletes and Untrained Older individuals," *Frontiers in Physiology,* August 30, 2019. On the cognitive side, see: Denise Park and Gerard Bischof, "The Aging Mind: Neuroplasticity in Response to Cognitive Training," *Dialogues in Clinical Neuroscience* 12, no. 1 (April 2022): 109–19. Finally, for the benefits of training body and brain together, see: Clemence Joubert and Hanna Chainay, "Aging Brain: The Effort of Combined Cognitive and Physical Training on Cognition as Compared to Cognitive and Physical Training Alone—A Systematic Review," *Clinical Interventions in Aging* 13 (July 20, 2018): 1267–1301.

236 use it or lose it across the board: Peter Reaburn and Ben Dascombe, "Endurance Performance in Masters Athletes," *European Review of Aging and Physical Activity* 5 (2008): 31–42; and Kevin Gries and Scott Trappe, "The Aging Athlete: Paradigm of Healthy Aging," *International Journal of Sports Medicine* 1055, no. 10, February 4, 2022. For a general overview of elite performance and aging, see Jeff Bercovici, *Play On* (Boston: Mariner Books, 2018); and Amanda Akkari et al., "Greater Progression of Athletic Performance in Older Masters Athletes," *Age and Aging* 44, no. 4 (July 2015). Also, Nick Heil, "Age Is Irrelevant When It Comes to Fitness," *Outside,* August 21, 2015. For an overview of science-based protocols for older athletes, see: Charlie Hoolihan, "Training Techniques for High Performance Masters Athletes," Ideafit.com, October 14, 2009.

236 Once you reach fifty: Abigail Barronian, "Here's How to Get Stronger after 50," *Outside,* July 12, 2018.

238 smart play works best: Jon Jachimowicz et al., "Why Grit Requires Perseverance and Passion to Positively Predict Performance," *Psychological and Cognitive Sciences* 115, no. 40 (September 17, 2018). Also see Ana Palis and Peter Quiros, "Adult Learning Principles and Presentation Pearls," *Middle East African Journal of Ophthalmology* 21, no. 2 (April–June, 2014): 114–22.

238 NYU neuropsychologist Elkhonon Goldberg: Elkhonon Goldberg, *The Wisdom Paradox* (New York: Gotham Books, 2005) Also see: Dilip Jeste et al., "The New Science of Practical Wisdom," *Perspectives in Biological Medicine* 62, no. 2 (2019): 216–36. Also a lot of Goldberg's thinking about how pattern recognition protects against cognitive decline comes from the study of cognitive reserve; see Jason Steffener and Yaakov Stern, "Exploring the Neural

Basis of Cognitive Reserve in Aging," *BBA: Molecular Basis of Disease* 1822, no. 3 (March 2012): 467–73.

241 the hippocampus is where: For neurogenesis in the adult human hippocampus, see: K. L. Spalding et al., "Dynamics of Hippocampal Neurogenesis in Adult Humans," *Cell* 153 (2015): 1219–27; and P. S. Eriksson et al., "Neurogenesis in the Adult Human Hippocampus," *Nature Medicine* 4 (1998): 1313–17; Rolf Knoth et al., "Murine Features of Neurogenesis in the Human Hippocampus Across the Lifespan from 0 to 100 Years," *PLoS One* 5, no. 1 (2010); Ashutosh Kumar et al., "Adult Neurogenesis in Humans: A Review of Basic Concepts, History, Current Research, and Clinical Implications," *Innovations in Clinical Neuroscience* 16, nos. 5–6 (May 1, 2019).

241 the easiest way to get the hippocampus to birth new neurons: For memory, emotion, and location, see: Bret Stetka, "Our Brain Is Better at Remembering Where to Find Brownies than Cherry Tomatoes," *Scientific American,* October 8, 2020; Jordana Cepelewicz, "The Brain Maps Out Ideas and Memories Like Spaces," *Quanta*, January 14, 2019. Also see Jennifer Talarico, Dorthe Berntsen, David Rubin, "Positive Emotions Enhance Recall of Peripheral Details," *Cognition and Emotion* 23, no. 2 (February 24, 2009): 380–98. For how this improves the aging brain, see: Francisco Mora, "Successful Brain Aging: Plasticity, Environmental Enrichment, and Lifestyle," *Dialogues in Clinical Neuroscience* 15, no. 1 (April 1, 2022): 45–52.

241 longevity hot spots: Deepan Dutta, "The Longevity Project/Part 1," *Summit Daily*, February 28, 2018; and Deepan Dutta, "The Longevity Project/Part 2," *Vail Daily*, February 13, 2018. Also see: https://en.wikipedia.org/wiki/List_of_U.S._counties_with_longest_life_expectancy.

# Index

# About the Author

STEVEN KOTLER is an award-winning journalist, a bestselling author, and the executive director of the Flow Research Collective. He is one of the world's leading experts on human performance. He is also the author of ten bestsellers (out of fourteen books), including the *New York Times* bestsellers *The Art of Impossible, The Rise of Superman, Abundance,* and *Bold.* His work has been nominated for two Pulitzer Prizes, translated into over fifty languages, and appeared in over a hundred publications, including the *New York Times Magazine,* the *Wall Street Journal, Atlantic, Wired,* and *Time.* Alongside his wife, author Joy Nicholson, he is the cofounder of Rancho de Chihuahua, a hospice care dog sanctuary.